Metagenomics - Basics, Methods and Applications

Edited by Wael N. Hozzein

Published in London, United Kingdom

IntechOpen

Supporting open minds since 2005

Metagenomics - Basics, Methods and Applications
http://dx.doi.org/10.5772/intechopen.78746
Edited by Wael N. Hozzein

Contributors
Takehiko Kenzaka, Cristina Cid, Eva Garcia-Lopez, Ana Moreno, Mario Tello, Alex Gonzalez, Gonzalo Gajardo, Natalia Valdes, Rodrigo Vargas, Joselin Rojas, Mick Parra, Krishnaraj P.U., Shrinivas Sabale, Padmaja Pralhad Suryawanshi, Raul Valle, Blancka Samaniego, Javier Apodaca, Leticia Arena, Wael Nabil Hozzein, Ahmed Mohammed Shuikan, Ayixon Sánchez-Reyes, Jorge L. Folch-Mallol

First published in London, United Kingdom, 2020 by IntechOpen
IntechOpen is the global imprint of INTECHOPEN LIMITED, registered in England and Wales, registration number: 11086078, 7th floor, 10 Lower Thames Street, London, EC3R 6AF, United Kingdom
Printed in Croatia

British Library Cataloguing-in-Publication Data
A catalogue record for this book is available from the British Library

Additional hard and PDF copies can be obtained from orders@intechopen.com

Metagenomics - Basics, Methods and Applications
Edited by Wael N. Hozzein
p. cm.
Print ISBN 978-1-83880-055-0
Online ISBN 978-1-83880-056-7
eBook (PDF) ISBN 978-1-83880-061-1

For EU product safety concerns:
IN TECH d.o.o., Prolaz Marije Krucifikse Kozulić 3, 51000 Rijeka, Croatia, info@intechopen.com or visit our website at intechopen.com.

Meet the editor

Prof. Wael N. Hozzein is a professor of Microbiology at the Faculty of Science, Beni-Suef University, Egypt. He received his PhD from Cairo University, Egypt in 2003, and then he worked as a visiting scientist at Newcastle University, UK; and Michigan State University, USA. Recently, he worked as the chair professor of the Bioproducts Research Chair, at King Saud University, Saudi Arabia. He has vast experience in the field of bacterial taxonomy with research interests in microbial biodiversity and applications. Prof. Hozzein is an author of more than 120 publications and he is a guest editor and a reviewer for several international journals. Additionally, he has been involved in many academic activities and educational reform projects and initiatives. He has been the principal investigator for several funded grants and has also received several awards among which was the State Encouragement Prize in Biological Sciences for 2015, Egypt.

Contents

Preface XI

Section 1
Basics of Metagenomics 1

Chapter 1 3
Introductory Chapter: Metagenomics and Metagenomic Approaches
by Wael N. Hozzein

Chapter 2 9
Soil Metagenomics: Concepts and Applications
by Shrinivas N. Sabale, Padmaja P. Suryawanshi and Krishnaraj P.U.

Section 2
Metagenomic Analysis 37

Chapter 3 39
High-Throughput Sequencing and Metagenomic Data Analysis
by Ahmed Shuikan, Sulaiman Ali Alharbi, Dalal Hussien M. Alkhalifah and Wael N. Hozzein

Chapter 4 53
The Use of Bioinformatic Tools in Symbiosis and Co-Evolution Studies
by Raúl Enrique Valle-Gough, Blancka Yesenia Samaniego-Gámez, Javier Eduardo Apodaca-Hernández and Maria Leticia Arena-Ortiz

Chapter 5 67
Metagenomics-Based Phylogeny and Phylogenomic
by Ayixon Sánchez-Reyes and Jorge Luis Folch-Mallol

Section 3
Applications of Metagenomics 83

Chapter 6 85
Microbial Community Structure and Metabolic Networks in Polar Glaciers
by Eva Garcia-Lopez, Ana Maria Moreno and Cristina Cid

Chapter 7 99
Dissemination of Intestinal Microbiota by Migratory Birds across Geographical Borders
by Takehiko Kenzaka

Chapter 8 115
Application of Metagenomics to Chilean Aquaculture
by Mario Tello, Natalia Valdes, Rodrigo Vargas, Joselin Rojas, Mick Parra, Gonzalo Gajardo and Alex Gonzalez

Preface

The advances in molecular biology, sequencing techniques, and data analysis tools in the past two decades have resulted in the evolution of metagenomics and enabled the study of uncultured organisms from different environments. Metagenomic analysis of the high-throughput sequences of a community sample has extended our understanding of the diversity, evolution, and functions of different organisms. This newly emerging science, which is one of the fastest developing fields in biology, is making important contributions to the fields of microbial ecology, bioremediation, discovery of industrial bioproducts, health, and many other areas.

This book is for the students starting their research projects in the field of metagenomics, for researchers interested in the new developments and applications in this field; and for teachers involved in teaching this subject. The book is divided into three sections as indicated from its title, namely; the basics of metagenomics, metagenomic analysis, and applications of metagenomics. It covers the basics of metagenomics from its history and background, to the analysis of metagenomic data as well as its recent applications in different fields. The book contains excellent texts at both the introductory and advanced levels, that describe the latest metagenomic approaches and applications, from sampling to data analysis for taxonomic, environmental, and medical studies.

The first chapter in this book, the introductory chapter, is a brief introduction on the basics of metagenomics, its background, and the metagenomic approaches. In the second chapter, the background on the application of different culture-independent techniques and metagenomics to study the microflora of the changing environment of soil is introduced.

It is very important for students and researchers to understand the methods they are going to use in their research. With that in mind, the next section is on methodologies and data analysis. In Chapter 3, an overview of the different sequencing techniques is presented in terms of principles, advantages, and limitations. In addition, third-generation sequencing technologies are also addressed. Chapter 4 focuses on different available bioinformatic tools that are suitable for studying symbiosis and coevolution processes in any given sample. Chapter 5 addresses the current methodologies to carry out community structure profiling, using single-copy markers and the small subunit of the rRNA gene to measure phylogenetic diversity from next-generation sequencing data. Interestingly, this chapter also presents a compendium of open source tools, easy to use by biologists and non-specialists to study microbial diversity in a phylogenetic context.

The last section of the book highlights the applications of metagenomics in different areas. Chapter 6 presents a complete analysis of the microbial diversity in the polar glacier habitats and an overview of the metabolic potentials and biogeochemical cycles in these interesting and less-studied ecosystems. In Chapter 7 the current state and the future implications of studying intestinal microbiota associated with migratory birds are described. The authors of this chapter discussed the use of next-generation sequencing to comprehensively analyze the intestinal microbiota of migratory birds, research on microbial communities traveling over long distances, that provides valuable information for the medical care, livestock industry, agriculture, and human health risks. Finally, in Chapter 8, the application of metagenomics as an emerging tool in aquaculture is discussed. This actual and potential application of metagenomics in aquaculture helps to understand the hosts-microbiota-pathogen-environment relationship underlying disease outbreaks and could be a key to develop strategies to improve productivity by increasing resistance to diseases and reducing the use of antibiotics with their negative environmental impact.

At the end, I would like to thank all the contributors for sharing their expertise in the field of metagenomics and for the appreciated academic integrity. I would like also to thank the reviewers for their precious time.

On behalf of all authors, I would like to warmly thank Ms. Sara Debeuc, the author service manager of this book, for her commitment, patience, and keen assistance which made this book possible. Our thanks are also extended to other staff of IntechOpen for their support.

Finally, the publication of this book was an interesting journey for me and I hope that the readers will enjoy reading it. I also hope this book will achieve its goals by contributing positively to the research community.

Wael N. Hozzein
Botany and Microbiology Department,
Faculty of Science,
Beni-Suef University,
Beni-Suef, Egypt

Section 1

Basics of Metagenomics

Chapter 1

Introductory Chapter: Metagenomics and Metagenomic Approaches

Wael N. Hozzein

1. Introduction

Metagenomics can be defined as the techniques and procedures that are used for the culture-independent analysis of the total genomic content of microorganisms living in a certain environment [1]. It has many useful applications with very promising potential in both medical and environmental microbiology. The most common use of metagenomics in environmental microbiology is studying the diversity of microbial communities in particular environments through the analysis of rRNA genes and how these communities change in response to changes in physical and chemical properties of these environments [2].

Metagenomics also provides an opportunity to obtain and identify novel enzymes with industrial applications from extreme environments where unculturable extremophiles live. In such circumstances, functional metagenomics enables the isolation of genes coding for extremozymes, enzymes that are capable of being catalytically active in extreme conditions, or genes that will allow for better understanding of the mechanisms that make such organisms resistant to extreme environmental conditions [3].

Metagenomics has special importance when it comes to studying soil microbiology. It is estimated that the number of distinct microorganisms in 1 gram of soil exceeds the number of microbial species cultured so far [4]. Therefore, metagenomics seems to be the ideal culture-independent technique for unraveling the biodiversity of soil microorganisms and to study how this biodiversity is affected with continuously changing conditions.

2. Sequencing technologies and metagenomics

Recently, taxonomic profiling, characterization, and analysis of microbial communities are being mostly performed using different next-generation sequencing (NGS) platforms. Metagenomic samples are high-throughput, short-read sequences, and the cost is relatively decreasing. In addition, these platforms are advantageous, avoiding the need for cloning of DNA fragments [5].

Recent advances in NGS technologies were developed to suit various numbers of applications, cost, and capabilities [6]. The most commonly used platforms are the 454 Life Sciences (Roche) and Illumina systems (Solexa) [7]. The 454 sequencing technology, which was the first commercially available next-generation technology, is based on the pyrosequencing technique. It provides high throughput and relatively cheap analysis [8]. During the sequencing reaction in this technique, nucleotide incorporation into the growing chain is detected by the capture of the released pyrophosphate, which is converted into a light through an enzymatic reaction. Different nucleotides are sequentially added into each nucleotide incorporation event; therefore the light signal can be attributed to a specific nucleotide. Finally, the light signals are converted into sequence information. In the 454 pyrosequencer, the DNA fragments are amplified after being fixed on beads in a water-oil emulsion [9]. Pyrosequencing has been employed widely in the analysis of microbial diversity in many environments including marine environments [10] and different soil environments [11, 12].

Illumina sequencing technology relies on the use of fluorescently labeled reversible terminator nucleotides. Instead of being chemically modified to prevent further DNA synthesis (dideoxynucleotides) which is the case with Sanger sequencing, the terminator nucleotides are attached with blocking group that can be removed from the nitrogen base in a single step. DNA synthesis takes place on a chip where primers are attached. After each cycle, the dyes attached to each nucleotide are excited by laser followed by scanning of the incorporated bases. In order for the next synthesis cycle to proceed, the blocking group and the dye are first removed by a chemical reaction. Illumina sequencing platform was successfully used to study microbial diversity in many environments [13–15].

In addition to the abovementioned technologies, recently developed sequencing technologies are available and being employed in metagenomic studies. These include SOLiD 5500 W Series developed by Applied Biosystems, single-molecule real-time (SMRT) DNA sequencing from Pacific Biosciences, and Ion Torrent semiconductor sequencing [7]. More innovative technologies are being developed that could be of great use for metagenomic studies in the near future. Strand sequencing technologies, currently being developed by Oxford Nanopore technologies, enable the sequencing of intact DNA strand that passes through a protein nanopore [16]. Irys Technology, developed by BioNano Genomics, represents one of the very promising new technologies in genomics era [7].

3. The metagenomic approaches

Metagenomics research strategy starts with selecting a proper ecological or biological environment of interest that hosts a wide variety of microbial communities which may have potential biotechnological applications. Environments that attract metagenomic researchers are mainly those characterized with extreme conditions or unique environmental conditions. These include environments with highly acidic or alkaline pH; high metal concentrations, pressures, or radiation; and high salinity or extreme temperatures [3].

Metagenomic analysis starts with isolating genomic DNA that represents the whole community in the soil sample, constructing a DNA library from the isolated DNA, and screening the available library for a target gene. It is important here to select the DNA extraction method that will provide enough yield and DNA that represents the diversity of the whole microbial community in the target environment. This is still one of the most challenging steps of metagenomic analysis. The chemical and physical characteristics of soils are very wide and complex, depending on the type of the soil examined, that will make it difficult to develop a reference method for DNA extraction from soils. Besides, soils contain many substances that are co-extracted with genomic DNA and harbor inhibitory effects on the downstream processing of the extracted DNA. Examples include humic and fulvic acids [17]. Therefore, optimization and comparison between different extraction methods are usually required for each type of soil [18–21].

A DNA library is then constructed from the genomic DNA isolated from the target environment. This is performed by fragmenting the isolated DNA into fragments with appropriate sizes that would allow for their cloning. This is performed by either using restriction enzyme digestion or mechanical shearing. DNA fragments obtained from such processes are cloned into the proper cloning vector. Plasmid vectors are used for small DNA fragments, and the libraries generated are called small-insert genomic libraries. Large inserts are cloning into cosmid or fosmid vectors which can hold inserts up to 40 kb in size or BAC vector which can carry inserts with sizes that exceed 40 Kb [22].

DNA libraries are usually constructed in a microorganism that is well-studied and is easy to manipulate inside the laboratory such as *Escherichia coli*. In case there is a need for expressing the genes within the DNA inserts in other microorganisms, shuttle vectors are used to transfer the libraries into a proper host [23].

Finally, a screening assay is applied to search for a gene of a particular function, and the gene product is functionally analyzed. There are two different metagenomic strategies that are commonly used in research. The first one is focused on the use of marker genes such as the ribosomal genes 16S rRNA [24] and 18S rRNA [25] to study the composition of the microbial community in a certain environment or specific protein-coding gene with medical or industrial importance [26–28]. Such a strategy is called targeted metagenomics. The second approach is the shotgun metagenomics, in which a wide coverage of genomic DNA sequences is achieved using high-throughput next-generation sequencing to assess the entire taxonomic structure or functional potential of microbial communities [29].

The most challenging aspect of the screening process in metagenomics is the analysis of a huge amount of sequence data that are generated from the constructed library. A wide range of bioinformatic tools has been developed over the years to help analyze the metagenomic data and compare it to available online databases.

Acknowledgements

This work was funded by the Deanship of Scientific Research at Princess Nourah bint Abdulrahman University, through the Research Groups Program Grant no. (RGP-1438-0006).

Author details

Wael N. Hozzein
Botany and Microbiology Department, Faculty of Science, Beni-Suef University, Beni-Suef, Egypt

*Address all correspondence to: hozzein29@yahoo.com

References

[1] Daniel R. The metagenomics of soil. Nature Reviews Microbiology. 2005; **3**(6):470

[2] Delmont TO, Robe P, Cecillon S, Clark IM, Constancias F, Simonet P, et al. Accessing the soil metagenome for studies of microbial diversity. Applied and Environmental Microbiology. 2011; **77**(4):1315-1324

[3] Mirete S, Morgante V, González-Pastor JE. Functional metagenomics of extreme environments. Current Opinion in Biotechnology. 2016;**38**: 143-149

[4] Torsvik V, Øvreås L. Microbial diversity and function in soil: From genes to ecosystems. Current Opinion in Microbiology. 2002;**5**(3):240-245

[5] Ansorge WJ. Next-generation DNA sequencing techniques. New Biotechnology. 2009;**25**(4):195-203

[6] Scholz MB, Lo CC, Chain PS. Next generation sequencing and bioinformatic bottlenecks: The current state of metagenomic data analysis. Current Opinion in Biotechnology. 2012;**23**(1):9-15

[7] Oulas A, Pavloudi C, Polymenakou P, Pavlopoulos GA, Papanikolaou N, Kotoulas G, et al. Metagenomics: Tools and insights for analyzing next-generation sequencing data derived from biodiversity studies. Bioinformatics and Biology Insights. 2015;**9**:BBI-S12462

[8] Ronaghi M. Pyrosequencing sheds light on DNA sequencing. Genome Research. 2001;**11**(1):3-11

[9] Goodwin S, McPherson JD, McCombie WR. Coming of age: Ten years of next-generation sequencing technologies. Nature Reviews Genetics. 2016;**17**(6):333

[10] Egge E, Bittner L, Andersen T, Audic S, de Vargas C, Edvardsen B. 454 pyrosequencing to describe microbial eukaryotic community composition, diversity and relative abundance: A test for marine haptophytes. PLoS One. 2013;**8**(9):e74371

[11] Wang X, Hu M, Xia Y, Wen X, Ding K. Pyrosequencing analysis of bacterial diversity in 14 wastewater treatment systems in China. Applied and Environmental Microbiology. 2012; **78**(19):7042-7047

[12] Alex A, Antunes A. Pyrosequencing characterization of the microbiota from Atlantic intertidal marine sponges reveals high microbial diversity and the lack of co-occurrence patterns. PLoS One. 2015;**10**(5):e0127455

[13] Fan W, Huo G, Li X, Yang L, Duan C, Wang T, et al. Diversity of the intestinal microbiota in different patterns of feeding infants by Illumina high-throughput sequencing. World Journal of Microbiology and Biotechnology. 2013;**29**(12):2365-2372

[14] Lentini V, Gugliandolo C, Bunk B, Overmann J, Maugeri TL. Diversity of prokaryotic community at a shallow marine hydrothermal site elucidated by Illumina sequencing technology. Current Microbiology. 2014;**69**(4): 457-466

[15] Hong C, Si Y, Xing Y, Li Y. IlluminaMiSeq sequencing investigation on the contrasting soil bacterial community structures in different iron mining areas. Environmental Science and Pollution Research. 2015;**22**(14): 10788-10799

[16] Jain M, Olsen HE, Paten B, Akeson M. The Oxford NanoporeMinION: Delivery of nanopore sequencing to the genomics community. Genome Biology. 2016;**17**(1):239

[17] Young JM, Rawlence NJ, Weyrich LS, Cooper A. Limitations and recommendations for successful DNA extraction from forensic soil samples: A review. Science & Justice. 2014;**54**(3): 238-244

[18] Finley SJ, Lorenco N, Mulle J, Robertson BK, Javan GT. Assessment of microbial DNA extraction methods of cadaver soil samples for criminal investigations. Australian Journal of Forensic Sciences. 2016;**48**(3):265-272

[19] Lim NY, Roco CA, Frostegård Å. Transparent DNA/RNA co-extraction workflow protocol suitable for inhibitor-rich environmental samples that focuses on complete DNA removal for transcriptomic analyses. Frontiers in Microbiology. 2016;**7**:1588

[20] Gupta P, Manjula A, Rajendhran J, Gunasekaran P, Vakhlu J. Comparison of metagenomic DNA extraction methods for soil sediments of high elevation Puga hot spring in Ladakh, India to explore bacterial diversity. Geomicrobiology Journal. 2017;**34**(4): 289-299

[21] Mazziotti M, Henry S, Laval-Gilly P, Bonnefoy A, Falla J. Comparison of two bacterial DNA extraction methods from non-polluted and polluted soils. Folia Microbiologica. 2018;**63**(1):85-92

[22] Simon C, Daniel R. Construction of small-insert and large-insert metagenomic libraries. In: Metagenomics. New York, NY: Humana Press; 2017. pp. 1-12

[23] Lam KN, Cheng J, Engel K, Neufeld JD, Charles TC. Current and future resources for functional metagenomics. Frontiers in Microbiology. 2015;**6**:1196

[24] Païssé S, Valle C, Servant F, Courtney M, Burcelin R, Amar J, et al. Comprehensive description of blood microbiome from healthy donors assessed by 16 S targeted metagenomic sequencing. Transfusion. 2016;**56**(5): 1138-1147

[25] West D. Use of an 18s rRNA metagenomics approach as a method of detection of multiple infections in field blood samples collected on FTA cards from cattle [Doctoral dissertation]. University of Salford; 2018

[26] Nurdiani D, Ito M, Maruyama T, Terahara T, Mori T, Ugawa S, et al. Analysis of bacterial xylose isomerase gene diversity using gene-targeted metagenomics. Journal of Bioscience and Bioengineering. 2015;**120**(2): 174-180

[27] Ufarte L, Laville E, Duquesne S, Potocki-Veronese G. Metagenomics for the discovery of pollutant degrading enzymes. Biotechnology Advances. 2015;**33**(8):1845-1854

[28] Lanza VF, Baquero F, Martínez JL, Ramos-Ruiz R, Gonzalez-Zorn B, Andremont A, et al. In-depth resistome analysis by targeted metagenomics. Microbiome. 2018;**6**(1):11

[29] Quince C, Walker AW, Simpson JT, Loman NJ, Segata N. Shotgun metagenomics, from sampling to analysis. Nature Biotechnology. 2017; **35**(9):833

Chapter 2

Soil Metagenomics: Concepts and Applications

Shrinivas N. Sabale, Padmaja P. Suryawanshi and Krishnaraj P.U.

"We know more about the movement of celestial bodies than about the soil underfoot."

—Leonardo Da Vinci, circa 1500s.

Abstract

Soil is a living entity of the Earth, and considered as one of the main reservoir of microbial diversity. Studying the soil microbial diversity is very much necessary, as they play an important role in maintaining the health of soil by recycling the nutrients, creating soil structure and humus. However, the culture dependent approaches fail to provide clear estimates of the diversity and untapped resources. Hence, study of the microbial diversity using culture independent approaches become necessary. The field of metagenomics helps in studying the genomes of the diverse soil organisms collectively in their natural habitat which holds the promising for accessing novel genetic resources. Application of the metagenomics to the soil environment is very challenging due to several difficulties; one of which is co-extraction of humic acid with nucleic acids which hinder downstream high throughout processes. However, applying sequencing methods to soil microbial communities will help in uncovering the hidden resources like novel genes, biomolecules and other valuable products which are yet to be discovered or still unknown. Different culture independent techniques and applications of the metagenomics to study the abundant microflora of the complex and changing environment of soil discussed herein.

Keywords: soil microbiome, soil health, metagenomics, DGGE, 16S rDNA, metagenome sequencing

1. Introduction

Life on earth is not only dependent on the celestial environment around the earth but it is also completely influenced by the soil under our feet. Earth is a pool of different complex ecosystems and the proper functioning of those ecosystems are very much required for sustenance and evolution of the life. Soil is an important ecosystem of the earth and essential for life and one of the most valuable resources available to us, which acts as a water filter, supports plant and animal life, source of the minerals and medicines. Soil is a living entity and existence of which is defined by the occurrence of the organisms in it. The formation of soil is not only dependent on the physical or chemical activity on the rocks, but also by the continuous activity of diverse microbial species which add or enhance the properties of soils in terms of formation of structure and function. Soil provides shelter to various soil harboring insects, reptiles, animals and a huge volume of microbes within the soil aggregates. It is the home for innumerable microbial species largely unnoticed and busy in maintaining the ecological balance of the earth. It is a complex and dynamic system, which directly or indirectly influences the food chain, nutrient cycles, and ecological equilibrium. Understanding the unexplored microbial world is high priority and is needed to come up with the ways of coping up with various climatic changes and for the welfare of every organism on the earth.

2. Soil health

Soil is an interconnected system with high levels of exchange of energy between organisms and physico-chemical components, which allows it to be a self-organized system. However, despite having the unique and incredible capacity to adapt to environmental changes, microbes are sensitive to land management and climate changes. Resistance is the capacity of the soil to maintain its health despite the magnitude of the change caused by any kind of perturbation. Resilience is the capacity of the system to return to its original state after a disturbance, which is also known as the self-healing capacity of the system (https://microbewiki.kenyon.edu/index.php/Soil_Health).

Soil has been neglected because it is termed as dirt. But from this dirt our ancestors learnt the skill of cultivation of crops and developed various cultural practices for the production of food. And our staple food started coming from the crops cultivated in the soil. Thus, agriculture—the science or art of farming, including cultivation of the soil for the growing of crops and the rearing of animals to provide food, wool, and other products came into existence (https://en.oxforddictionaries.com/definition/agriculture). Being a most diverse active environment with huge number of known and unknown microbial species, soil becomes a challenging ecological niche for study. The biochemistry of soil reflects several unknown functions which are very much important for sustenance of life. However, the recent technologies using heavy machinery and management practices intensified agriculture that resulted in extensive cultivation without enriching the soil. This overuse has initiated the deterioration of the cultivable farm lands through loss of fertility, vigor, soil structure and capacity to sustain life. In short, many cultivable areas have turned saline or uncultivable. Agricultural land is simultaneously being lost to non-agricultural uses. In addition, the intensive use of land in production through multiple cropping, reduced fallow periods, excessive use of agrochemicals, spread of monocultures are components of management is perceived as leading to land degradation and the undermining of its long-term productive potential (www.fao.org).

The formation of 1 cm of topsoil in its natural course requires 300 years but the resultant loss of soil due to erosion is high (Soil loss in India was 16.4 t/ha/year as reported in State of Indian Agriculture Report, 2015–2016). The population around the globe has been expanding rapidly and standing at around 7.3 billion in 2016 and still counting. With the increase in the population number of challenges around global sustainability also increases, including the need for more food and space. The demand and supply of food is the major concern for all the countries and it can be noted that we are falling short of cultivable land due to soil erosion, saline soils, excessive use of pesticides, herbicides, inorganic fertilizers and the shift of land use to housing sector. According to the soil scientist Dr. Elaine Ingham, "If we lose both bacteria and fungi, then the soil degrades".

Overuse of chemical fertilizers and pesticides have effects on the soil organisms that are similar to human overuse of antibiotics [1]. Therefore, sustaining and improving the characteristics of soil will be an utmost priority for generations to come and we have to focus on keeping our soils alive by maintaining the cultivable land or rejuvenating the barren lands. Thus, soil health becomes one of the most important factors in terms of agriculture and forest ecosystems and very much necessary for survival of living beings.

2.1 Importance of soil microorganisms

Biological fertility of the soil comprises of the soil microbes which greatly influence the structure and functioning of the soil ecosystem. The maintenance of the physical fertility and chemical fertility is driven by the metabolic repertoire of the soil microorganisms. Soil microbes are part of a very complex food web that occurs in soil. Soil biota is indispensable for key soil function such as decomposition of soil organic matter, nutrient cycling and formation of soil aggregates and billions of microbes reside in a single gram of soil [2]. The most numerous microbes in soil are the bacteria, followed in decreasing numerical order the fungi, soil algae and soil protozoa. The work that is turned out by bacteria does benefit the plants and other living organisms. Plants are unable to absorb the nutrition from the soil without microbes working in the soil. Although microbes tend to receive more attention for the diseases they cause, they are likely to be more significant for their role in maintaining the health of the plants by protecting them from other microorganisms, providing vitamins, nutrients and influencing the developmental processes [3]. The microbial activity in the rhizosphere is essential for plant functioning as it assists the plant in nutrient uptake and offers protection against pathogen attack [4]. Rhizosphere, a region of soil which is influenced by the plant roots, a microenvironment where a great microbial diversity thrives in close association with plant roots where in various abiotic and biotic interactions take place. Abiotic factors influence the structure of microbial communities which have evolved mechanisms to deal with occupation of the space that allows them to obtain nutrients excreted by the plant roots. The organism diversity of the bulk-soil or the rhizospheric soil of different plants has been extensively studied; most studies have reported a wide range of organisms from those soil portions [5]. Surprisingly, a soil sample could contain up to 4×10^6 different taxa [6]. This suggests that soil is a large reservoir for the discovery of several compounds that may have applications in agriculture, human health or industry. The relevant insights of the studies related to soil are that each of them has shown different aspects to study and analyze. In some cases, they have provided innovative views of the microbial ecology on extreme environments for the development of life. At the same time, others have found novel biocatalysts, new antibiotics, personalized medicine, bioremediation and other potential applications in biotech industry [7].

Microbiological studies in the soil environment are hampered by the fact that the largest proportion of soil bacteria as yet cannot be cultured [8], and the microbes found using common culture methods are rarely abundant in any environmental niche from which they were cultured. The microbes isolated from the soil mostly belong to the four major phyla namely *Proteobacteria*, *Actinobacteria*, *Firmicutes* and *Bacteroidetes*, as those minute organisms are easily cultivable under laboratory conditions. Such information suggests that there is the need of molecular techniques which overcomes the need to isolate and study each organism for in depth characterization. However with the development in metagenomics a more complete picture of the rhizosphere microbiome can be unearthed [9] and many new microbial players in the rhizosphere are on their way to be discovered.

3. Soil metagenomics: a concept

Even though soil microbes dominate the biosphere, most microbes in nature have not been studied because the standard culturing techniques could help in the study of those organisms which can be cultured on artificial media and rest of the organisms are still unknown since lack of the universal artificial media and growth conditions. Instead of cultivating different organisms in pure form and studying the morphology and characteristics, metagenomics, studies all the organisms present in the environmental sample at once by the combining traditional microbiology and molecular biology. Microbial diversity is the major driving force of fundamental metabolic processes in dynamic environment like soil and a number of microbial species are associated with the plant rhizosphere. Therefore, a basic understanding of diversity of soil biota is required in order to preserve the integrity, function and long term sustainability of natural and managed terrestrial ecosystems. The opportunity that stands before microbiologists today is akin to a reinvention of the microscope in the expanse of research questions it opens to investigation [10].

Metagenomics provides a new way of examining the microbial world that not only will transform modern microbiology but has the potential to revolutionize understanding of the entire living world and their functions in different ecosystem. In metagenomics, the power of genomic analysis is applied to entire communities of microbes, bypassing the need to isolate and culture individual bacterial community members in order to identify the microbes in the community. The new approach and its attendant technologies will bring to light the myriad capabilities of microbial communities that drive the planet's energy and nutrient cycles, maintain the health of its inhabitants, and shape the evolution of life. Metagenomics is expected to generate knowledge of microbial interactions so that they can be harnessed to improve human health, food security, and energy production.

Metagenomics combines the power of genomics, bioinformatics, and systems biology. Operationally, it is novel in that it involves study of the genomes of many organisms simultaneously. It provides new access to the microbial world. Although community ecology is not new to microbiology, the power of genomics in the study of communities brings in an unparalled opportunity.

Meta in the first sense means that this new science seeks to understand biology at the aggregate level, transcending the individual organism to focus on the genes in the community and how genes might influence each other's activities in serving collective functions. In the second sense, meta also recognizes the need to develop computational methods that maximize understanding of the genetic composition and activities of communities so complex that they can only be sampled, never completely characterized.

Metagenomics, still a very new science, has already produced a wealth of knowledge about the uncultured microbial world because of its radically new ways of understanding microbial world. All metagenomics studies take the same first step: DNA is extracted directly from all the microbes living in a particular environment. The mixed sample of DNA can then be analyzed directly, or cloned into a form maintainable in laboratory bacteria, creating a library that contains the genomes of all the microbes found in that environment.

Since the invention of the simple microscope by Anton Von Leeuwenhoek we are able to identify and study only 1% of the total microbial diversity available on Earth. The rest of the 99% of the microbes are not easily accessible to us as they are busy with their life cycle and maintenance of the microbial balance of our planet. Our current knowledge about the culturable microbial techniques is unable to achieve the complete picture of the diversity and functions of the micro-organisms residing even in a single environment. Thus the development of new techniques holds huge potential to overcome the difficulties to capture complete microbial profile.

The study of microbial diversity is very much important in a particular habitat as the microbes helps in maintenance or modification of the environment and it is also responsible for survival of surrounding organisms and also helps in the process of evolution of the other living organisms. It is therefore, time to investigate the intricate relationship between the soil microbiota and plants and how this relationship benefits both of the organisms in spite of the environmental changes.

In 1904, Lorenz Hiltner coined the term, "rhizosphere"—a word partly originated from the Greek word "rhiza", meaning root which describes the plant-root interface [11]. Hiltner described the rhizosphere as the area around a plant root that is inhabited by a unique population of microorganisms. But in recent years based on the relative proximity and influence to root the rhizosphere is refined to include three zones. The endorhizosphere includes cortex and endodermis in which microbes occur in very close association with the roots of the plant; they are present inside the roots and referred to as endophyte at times. They are resultant of plant root exudates which attract them to colonize the roots. The rhizoplane is the medial zone directly adjacent to the root including the root epidermis and mucilage. The outermost zone is the ectorhizosphere which extends from the rhizoplane out into the bulk soil and it is the house to the organisms which are either free living or non-symbionts which are highly influenced by management or cultural practices. The rhizosphere is not a region of definable size or shape, but consists of a gradient in chemical, biological and physical properties which change both radially and longitudinally along the roots.

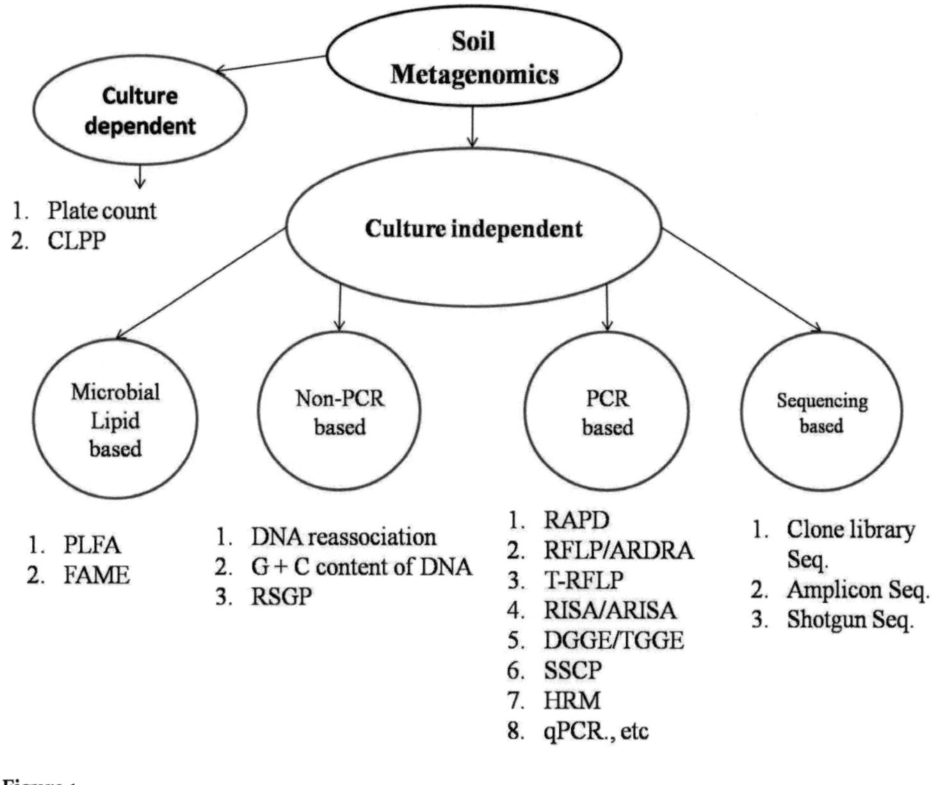

Figure 1.
Methods for studying microbial diversity harboring in the soil.

Discovering the microbiota of the soil environment requires selection of the suitable method(s) for capturing the maximum diversity at a particular time and space. It is obvious that based on a particular interest of the researcher, the approach to understand microbial diversity may be different from others or same. To determine spatial heterogeneity, considerable amount of work is still needed, for example, how representative a 0.1 mg sample of soil is with respect to the larger environment from which it was taken. Selection of proper method for extracting the meaningful information is very much important to escape the biasness of the results or to reduce the error rate in estimating the diversity by using statistical significance as well. To understand the overall microbial diversity, the researchers use one or combination of the methods given below for collecting the maximum information using their own standard sample size (**Figure 1**).

4. Culture dependent approaches

4.1 Plate count

Standard culture technique involves selective plating, direct counting viable colonies/cells and characterization of microorganisms on artificial or synthetic growth media in laboratory condition [12]. The major limitation of the culture dependent techniques is that we are unable to study the microbes which cannot be grown on artificial media. In spite of putting efforts in improving the culture media which can copy the natural conditions for the growth of the microbes, it was seen that the majority of the fraction of microbes were still unculturable. The minimal media, culture dilutions and the extended incubations helps in recovery of few slow growing microbial species which were considered as uncultivable earlier, but the abundance of the microbes in soil, which can be seen under microscope, still remains untapped or untouched. A technique has been devised for the cultivation of uncultured microorganisms from different environments that involved encapsulation of cells in gel micro-droplets for large-scale microbial cultivation under low nutrient flux conditions [13].

4.2 Community level physiological profiling (CLPP)

Garland and Mills developed a technique to assess the potential functional diversity of the bacterial population through Sole Source Carbon Utilization (SSCU) patterns [14]. Based on carbon utilization, BIOLOG (https://biolog.com) introduced the technique into ecological studies to estimate metabolic potential of microbial communities. The BIOLOG plate method was used for comparing metabolic activity of heterotrophic microbial communities from different habitats using redox chemistry. These plastic microtiter plates contain colorless tetrazolium dye with 95 different carbon substrates (e.g. carbohydrates, amino acids, amines, amides, carboxylic acids, etc.) in wells, and one well without any substrate used as control. Soil samples are shaken in a suitable solution; the soil solution is inoculated into the plate wells and incubated at constant temperature. Soil microorganisms oxidize the substrates in the microtiter plate wells and also reduce the tetrazolium dye to a violet formazan which is measured spectrophotometrically [15]. Currently BIOLOG has developed various microtiter plates for identification of aerobes, anaerobes, yeasts and filamentous fungi. This method is rapid, convenient and can help to identify over 2900 bacteria, fungi and yeast. But the drawback is the microbes which are cultivable and able to grow in high-nutrient conditions can only be studied. This method was used mainly to investigate microbial communities from bulk and rhizosphere soil and to estimate the impact of stressors on soil microbial communities [15]. The BIOLOG microplates were used for identifying the bacterial population in the contaminated soils around the Explosive Institute of Riyadh community (Saudi Arabia) which shown *Proteobacteria* to be the dominant phylum in all the five regions with Burkholderiaceae, Chromatiaceae and Pseudomonadaceae common and dominant families representing 49% of the relative abundance at the family level [16]. Compared to earlier microtiter plates, BIOLOG EcoPlate are more in use for Community Level Physiological Profiling (CLPP) as it contains three replicates of 31 different carbon sources and one control well per replicate (Biolog Inc., Hayward, CA, USA) [15].

5. Culture independent approaches

The viable source of information regarding the microbial players in the soil can be discovered through the biomolecules such as lipids, DNA, RNA and proteins. The extraction procedures of the biomolecules from the soil are another challenge as the content of the soil, structure and humic acids varies from place to place, time to time. Over the years, several procedures has been devised for extracting the biomolecules such as nucleic acids but still there is the compromise in concentration or quality of the biomolecules. However, in recent years there has been huge progress in developing the new techniques for studying the microbes in soil and other environments.

5.1 Microbial lipid based techniques

As we know fatty acids are the components of the cellular membrane of all living cells, and their composition can reveal the types of organisms present without actual culturing the micro-organisms. The constant proportion of fatty acids microbial cell biomass as well as the presence of signature fatty acid and their profile helps to differentiate major taxonomic group within the community.

5.1.1 Phospholipid fatty acid analysis (PLFA)

Phospholipid fatty acid analysis (PLFA) quantifies a set of biomarkers that track primarily viable biomass, avoids culturing of micro-organisms, and represents in-situ conditions. PLFA provides a community measurement that is phenotypic rather than genotypic in nature. It does not give information on species composition but rather is analogous to the ecological concept of functional groups [17]. In a comparison study regarding the PLFA analysis and 16S rRNA gene metabarcoding of bacterial communities across the biomes, the PLFA profiling has been found better in distinguishing bacterial community [18]. It was also noticed that the PLFA profiling was better at detecting community responses to heavy metal pollution [19]. The other method which is also based on fatty acid extraction and profiling is fatty acid methyl ester and was used for estimating microbial biomass and characterizing microbial community composition in soil [20].

5.1.2 Fatty acid methyl ester (FAME)

Fatty acid methyl ester (FAME) provides information on the microbial community composition based on groupings of fatty acids. The fatty acids are extracted by saponification followed by derivatization to give the respective FAMEs, which are then analyzed by gas chromatography. The pattern thus obtained is compared to a reference FAME database to identify the fatty acids and their corresponding microbial signatures by multivariate statistical analyses [21]. Bacterial fatty acids are highly conserved due to their role in cell structure and function and are the major constituents of the lipid bilayer of bacterial membranes and lipopolysaccharides. They have been used extensively for taxonomic and identification purposes. Whole cellular FAME content is a bacterial profile and is a direct and stable expression of the cellular genome. The cellular fatty acid pattern is a phenotypic character that is not affected by mutations, acquisition or loss of plasmids. The use of fatty acid analysis by gas chromatography for the identification of bacteria is rapid, efficient, reproducible and used for the identification of both clinical and environmental isolates. It has been used to study microbial community composition and population changes due to agricultural practices. Miura and her associates reported that EL- FAME method was simple and would produce similar results to PLFA method for bacteria in both quantitative and qualitative assessments when comparing different soils across ecosystems [20].

The importance of FAME analysis for the identification of bacteria is based on the large structural differences within these molecules viz., (i) variation in length (8–20 C-atoms), (ii) presence of saturated and monounsaturated fatty acids, (iii) occurrence of branched fatty acids (iso and anteiso fatty acids or methylated within the molecule), (iv) occurrence of cyclopropane fatty acids (17:0c, 19:0c), (v) occurrence of hydroxy-fatty acids with an OH-group at position two or three of the molecule. For classification or identification of bacteria the presence of distinct fatty acids and their relative amount is analyzed and compared with the fatty acid profiles of reference strains [22].

The microbial community characterization using nucleic acids has been further discussed as follows:

5.2 Non-PCR based techniques

5.2.1 DNA re-association

DNA re-association kinetics measures the genetic complexity of the microbial community and has been used to estimate microbial diversity. Total DNA is extracted from environmental samples, purified, denatured and allowed to re-anneal. The rate of hybridization or re-association will depend on the similarity of sequences present. As the complexity or diversity of DNA sequences increases, the rate of re-association of DNA will decrease. Under specific conditions, the time needed for half of the DNA to re-associate (the half association value $C_0t_{1/2}$) can be used as a diversity index, as it takes into account both the amount and distribution of DNA re-association. The parameter controlling the re-association reaction is concentration of DNA product (C_0) and time of incubation (t), usually described as the half association value, $C_0t_{1/2}$ (the time needed for half of the DNA to re-associate). Under specific conditions, $C_0t_{1/2}$ can be used as a diversity index, as it takes into account both the amount and distribution of DNA re-association [23]. Alternatively, the similarity between communities of two different samples can be studied by measuring the degree of similarity of DNA through hybridization kinetics [24].

5.2.2 Guanine plus cytosine (G+C) content of DNA

Differences in the guanine plus cytosine (G+C) content of DNA can be used to study the bacterial diversity of soil communities [25]. It is based on the knowledge that microorganisms differ in their G+C content and that taxonomically related groups only differ between 3 and 5%. Even though GC fractionation provides coarse level of resolution as different taxonomic groups may share the same G+C range, it is probably the only technique that is completely independent of any previous knowledge regarding which bacterial populations comprise the community or their genomic content [26]. However, this method can be used with other PCR based methods like DGGE or TGGE for better accessibility of the microbial picture.

5.2.3 Reverse sample genome probing (RSGP)

This method utilizes genome microarrays to analyze microbial community composition. RSGP has four steps: (1) isolation of genomic DNA from pure cultures; (2) cross-hybridization testing to define species with less cross hybridization (<70%). DNA fragments with greater than 70% cross-hybridization are considered to be the same species; (3) preparation of genome arrays by spotting known amounts of denatured genomic DNAs from all identified standards onto a solid support; and (4) random labeling of a defined mixture of total community DNA and internal standard, hybridization of the labeled probe with the genome array and detection and analysis of the individual dot hybridization data [27]. Due to low level of hybridization, low levels of gene expression cannot be quantitated which limits the use of this technique. However, this technique allows the coverage of the uncultured component of environmental microbial communities. Although possible in principle, the problem of linking a specific DNA fragment to a particular strain is formidable and requires extensive characterization of any metagenome through cloning and sequencing.

5.3 PCR based methods

Targeting the 16S rDNA is used extensively to study prokaryote diversity and allows identification of prokaryotes. The prediction of phylogenetic 18S rDNA and internal transcribed spacer (ITS) regions are increasingly used to study fungal communities. Soil community DNA is extracted, purified and amplified using either specific or universal primers, the resultant products are then separated by various ways and analyzed accordingly.

5.3.1 Highly repeated sequence characterization or microsatellite regions

During the process of evolution both in prokaryotic and eukaryotic organisms, there is the accumulation of highly repetitive short DNA sequences (1–10 bp) throughout their genomes, which can be used in differentiation between the organisms at species or strain level. Highly repeated sequences are also termed as microsatellite regions and have been used for identification of mycorrhiza. Fingerprints of the PCR-amplified microsatellites can be compared using similarity indices to investigate differences between or among the species. The design of primers is solemnly dependent on the sequence information of microsatellite regions. The use of this method to study microbial diversity may be limited depending on the complexity of the community [12].

5.3.2 Random amplified polymorphic DNA

In 1990, William and his team developed a method which includes amplification of DNA fragments by using short arbitrary primer targeting multiple loci in genomic DNA, generating unique profile (amplicons of various lengths [28]). Both genomic variations between bacterial species and genetic polymorphism between bacterial strains could be identified based on the differences in the molecular size and the number of DNA fragments obtained (**Figure 2**). RAPD analysis was used to study metagenome diversity in soil microbial community of arid zone plants [29] and in soil affected by industrial pollutants [30].

5.3.3 Restriction fragment length polymorphism (RFLP)/amplified ribosomal DNA restriction analysis (ARDRA)

It is another tool used to study microbial diversity that relies on DNA polymorphisms. PCR amplified rDNA is digested with a 4-base pair cutting restriction enzyme. Different fragment lengths are detected using agarose or non-denaturing polyacrylamide gel electrophoresis in the case of community analysis. RFLP fingerprint can be used to measure bacterial community structure. ARDRA is a DNA fingerprinting technique based on PCR amplification of 16S ribosomal DNA using primers for conserved regions, followed by restriction enzyme digestions and agarose gel electrophoresis. ARDRA was used successfully to study and compare the microbial diversity in copper contaminated soils [31]. Sklarz and his associates evaluated the use of amplified rDNA restriction analysis assay for identification of bacterial communities and concluded that ARDRA based dendrograms may not mirror 16S rDNA sequence based phylogenetic trees [32].

5.3.4 Terminal restriction fragment length polymorphism (T-RFLP)

T-RFLP follows the same principle as RFLP except that one PCR primer is labeled with a fluorescent dye, such as TET (4,7,2V,7V-tetrachloro-6-carboxyfluorescein) or 6-FAM (phosphoramidite fluorochrome 5-carboxyfluorescein), which allows detection of only the labeled terminal restriction fragment [33]. This simplifies the banding pattern, allowing the analysis of complex communities as well as providing information on diversity as each visible band represents a single operational taxonomic unit or ribotype. The banding pattern can be used to measure species richness and evenness as well as similarities between samples. Terminal restriction fragment length polymorphism (T-RFLP) has higher resolution and is more comprehensive than cultivation-based methods [13]. This procedure can be automated to allow sampling and analysis of a large number of soil samples with recent developments in bioinformatics, several Web-based T-RFLP analysis programs have been developed, which enable researchers to rapidly assign putative identities based on a database of fragments produced by known 16S rDNA

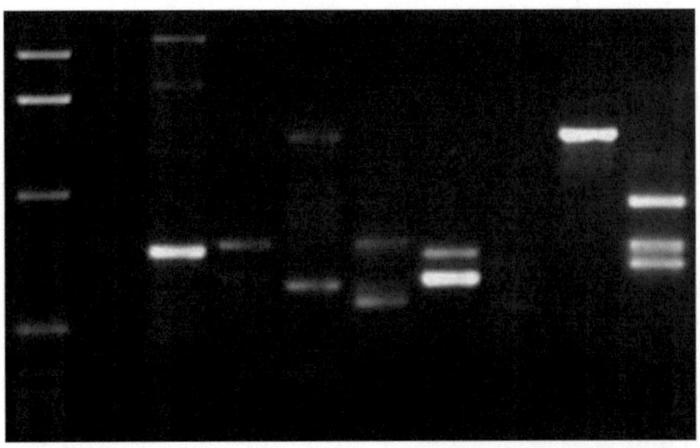

Figure 2.
RAPD profile of bulk soil DNA.

sequences. This technique has been successfully applied to the composition and diversity analysis of soil microbial communities under different environmental conditions [34–36]. T-RFLP analysis was used to reveal the composition and diversity of soil bacterial communities along an altitude gradient in Wuyi Mountains addressing relationship between the composition and structure of the soil microbial communities and the vegetation types and cause of the difference in soil microbial communities under different vegetation types [13].

5.3.5 Ribosomal intergenic spacer analysis (RISA)/automated ribosomal intergenic spacer analysis (ARISA)

Ribosomal intergenic spacer analysis (RISA) is a simple, single-step PCR-based method for profiling microbial diversity that detects the variation in size of the intergenic transcribed spacer (ITS) region between the bacterial 16S and 23S rRNA genes [37]. RISA has been extensively used to profile microbial diversity in a range of environmental niche. Intergenic spacer (IGS) region between the 16S and 23S ribosomal subunits is amplified by PCR, denatured and separated on a polyacrylamide gel under denaturing conditions. This region encodes tRNAs and is useful for differentiating between bacterial strains and closely related species because of heterogeneity of the IGS length and sequence. In RISA, the sequence polymorphisms are detected using silver stain while in ARISA the forward primer is fluorescently labeled and is automatically detected using laser. ARISA is a rapid and effective community analysis technique which can be used in conjunction with other accurate labor-intensive methods (e.g., 16S rRNA gene cloning and sequencing) for fine-scale spatial and temporal resolution [38]. Delmont and co-workers evaluated the different DNA extraction protocols by using RISA in study of soil microbial communities and reported that the total community DNA extracted from different DNA extraction procedures generated the different RISA profiles [39].

5.3.6 Temperature/denaturing gradient gel electrophoresis (TGGE/DGGE)

Muyzer with his teammates expanded the use of DGGE to study microbial genetic diversity [40]. In DGGE, DNA is extracted from soil samples and amplified using PCR with universal primers targeting part of the 16S or 18S rRNA sequences. The 5′-end of the forward primer contains a 40 base pair (16S rRNA) or 50 base pair (18S rRNA) GC clamp to ensure that at least part of the DNA remains intact or to avoid the complete dissociation of the amplified products into single strands which might flow away from the gel. This is necessary as the separation of amplified DNA on a polyacrylamide gel with a gradient of increasing concentration of denaturants (formamide and urea) will occur based on melting behavior of the double-stranded DNA (**Figure 3**). TGGE uses the same principle as DGGE except the gradient is temperature rather than chemical denaturants. Polymorphism based on the separation of partially melted 16S rDNA a linear temperature gradient. It represents the sequence variations other than the restriction sites also. Sequence variation among different PCR amplicons determines the melting behavior, and therefore amplicons with different sequences stop migrating at different positions in the gel. However, it covers only less than less than 400 bp of 16S gene. Conservative fragments of available 16S rDNA sequences were mined and searched for candidate primers within the fragments by measuring the coverage rate defined as the percentage of bacterial sequences containing the target. Thirty predicted primers with a high coverage rate (>90%) were identified [41] and can be successfully used for generating DGGE fingerprints. Abundance of denitrifying genes and microbial community structure in volcanic soils [42], assessment of silver nanoparticles on soil bacterial diversity [43],

Figure 3.
DGGE profile of the 16S rDNA (V5 region) of soil rhizosphere samples.

effect of long term fertilization on bacterial and fungal diversity in brown soils [44], Changes of Soil Bacterial Diversity in a Semi-Arid Ecosystem [45] has been successfully studied by using DGGE profiles.

How to get more out of molecular fingerprints remains the question, because estimating the species diversity is the most important factor towards better understanding of the microbial load in the given soil environment. Various investigations indicate that the species richness is the simplest measure of diversity. Any characterized soil environment shows few common species but in greater abundance as compared to more uncommon species harboring in the same environment but in less number therefore, one should consider the species evenness also. A common point of agreement on the diversity is that, species richness and evenness aggregately estimates the diversity and these components should be defined so that they are independent of each other [46]. One of the most commonly used evenness measure is Pielou's evenness index, evenness expresses how evenly the individuals in the community are distributed over the different species [47].

Claude Shannon originally proposed this measure which has been useful in comparing diversity between the different habitats [48]. Shannon index is easy for calculation and interpretation, the Shannon index generally ranges between 1.5 and 3.5 for many ecological habitats. Simpson diversity is less sensitive to richness and more sensitive to evenness; whereas Shannon diversity is more sensitive to evenness. For comparing the similarity between the samples, one can apply Jaccard index or Sorensen index. They are most widely used and are based on the presence/absence of species in paired assemblages and are simple to compute [49]. A modified version of Jaccards has been proposed by Chao and his colleagues to include the effect of unseen shared species, based on either (replicated) incidence or abundance based sample data [50].

As the soils have a dynamic nature, there is always a shift in microbial population during the different time or during different seasons. At a given time, particular number of species significantly dominates over other and vice versa. Therefore, such shift in microbial load can be estimated using different comparison tools. Some of them are cluster analysis [51], moving window analysis [52] or by visual inspection [53] or the Dice index [54] which can be applied for estimation of microbial shift during a defined period of time. The percent change in microbial composition between the two sampling interval can be calculated by subtracting percent similarity (calculated by any of the similarity indices) from 100. This can be done for consecutive sampling points over a period of experimentation. Using that % change values, moving window analysis is plotted between consecutive sampling points. The rate of change (Δt) value is calculated as the average of the respective moving window curve data points [43]. Higher Δt values represent higher shifts between two successive sampling points.

The above diversity analysis parameters are based on the relative abundance of the species in the given sample and extensively used to analyze denaturing gradient gel electrophoresis profiles.

5.3.7 DNA microarray

Nucleic acid (DNA/RNA) hybridization using specific probes is another qualitative and quantitative tool in molecular bacterial ecology. Oligonucleotide or polynucleotide probes designed from known sequences ranging in specificity from domain to species can be tagged with markers (fluorescent) at the 5′-end [55]. Based on nucleic hybridization, DNA microarrays are developed and are used to detect and identify bacterial species in soil or any other environmental samples. In this method, a single chip contains thousands of probes with high specificity can be used in identifying microbial species in soil or environmental samples. The amplified products from the soil DNA is hybridized against the known molecular probes, which are attached on the microarrays. The hybridized spots are detected and scored using microscopy. Its advantage is the rapid and replicated evaluation of the samples [56] but while analyzing the soil or environmental sample there may be risk of cross hybridization. Further, there are two different microarray chips namely 16S rRNA gene microarray or PhyloChip and Functional gene microarray or GeoChip (**Figure 4**).

PhyloChip is the most widely used phylogenetic array. In here, 16S rRNA genes are extracted followed by PCR amplification of metagenome DNA and then biotin labeled for PhyloChip hybridization, the signals are detected using digital image detection. It is Affymetrix-based technology consisting 25-mer oligonucleotide probes which differentiate between the 16S rRNA gene sequences in microbial communities. The recent version of the PhyloChip (G3) has probes targeting ~60,000 operational taxonomic units (OTUs), representing two domains (Archaea and Bacteria), 147 phyla, 1123 classes, 1219 orders, 1464

Figure 4.
DNA microarray chip (PhyloChip or GeoChip).

families, and 10,993 subfamilies [57]. Functional gene arrays are used in targeting genes involved in various biogeochemical cycling processes which help in determining the functional composition microbial communities. GeoChip, a functional gene array is widely used which targets hundreds of functional genes involved in biogeochemical, ecological, and environmental analyses. Arrays for detecting specific functional processes, such as nitrogen cycling, methanotrophy, stress responses, hydrogen activity etc. are available. The version 5.0 GeoChip contains about 167,000 50-mer oligonucleotide probes covering ~395,000 coding sequences from >1590 functional genes related to various microbes, mineral cycling, energy metabolism, antibiotic resistance, metal homeostasis and resistance, secondary metabolism, organic remediation, stress responses, bacteriophages, and virulence [58].

5.3.8 Single-strand conformation polymorphism (SSCP)

SSCP was developed to detect gene polymorphism in human DNA and mutations by comparing PCR products [59]. DNA fragments are amplified, followed by denaturation and separation in non-denaturing polyacrylamide gel. Single-stranded DNA is separated on a polyacrylamide gel based on differences in mobility caused by their folded secondary structure. DNA fragments of equal size with no denaturant results into folding and movement, depending on the DNA sequences. Single stranded DNA forms secondary structures which are unique to its nucleotide sequence. This secondary structure hinders the movement of DNA in polyacrylamide gel and hence, the banding pattern is obtained at different position from the fragment of equal size but amplified from genetically different species. The advantage of SSCP over DGGE is that here GC clamp is not required during PCR but has a limitation with that there is high rate of re-annealing of DNA strands after an initial denaturation during electrophoresis which can however be overcome using a phosphorylated primer during PCR [21]. Smalla and her team studied the bacterial diversity of the soils by assessing the DGGE, T-RFLP, and SSCP fingerprints of 16s rRNA gene fragments. Although, the fragments amplified comprised different variable regions and lengths, DGGE, T-RFLP and SSCP analyses led to similar findings: (a) a clustering of fingerprints which correlated with soil physico-chemical properties, (b) little variability between the replicates of the same soil, (c) the patterns of the two brown soils were more similar to each other than to those of the other two soils, and (d) the fingerprints of the different soil types revealed significant differences in a permutation test [60]. SSCP fingerprints were also used in study of microbial diversity in landslide soils; the more detailed profile of fungal diversity was obtained using this method [59].

5.3.9 High resolution melt curve (HRM)

High resolution melt curve typically involve qPCR amplification followed by a melting curve collected using a fluorescent dye. It follows extraction of nucleic acids and qPCR with fluorescent labeled dye. Following amplification and quantification of the 16S rRNA gene, a high-resolution melting curve analysis is obtained. The procedure melts the amplified 16S rRNA gene products between the temperatures ranging from 72°C and ending at 95°C, with fluorescence measurements taken at every 0.1°C increment. Melting curves are normalized to relative fluorescence units (RFU) in a specified "melt region" (e.g., 83.5–89.5°C), thereby negating the effect of absolute RFU values. The melt region is autocalled by the melt analysis software (Precision Melt Analysis) [61]. The effects of herbicide on soil bacterial diversity were efficiently studied using HRM analysis [62].

5.3.10 Real time PCR (qPCR)

Real time PCR (qPCR) helps in real time quantification of the active population present in the environmental sample by targeting ribosomal genes specific to different species of the microbial community. It gives an idea of phylogenetic community composition by assessing the range of phyla or classes [63]. Quantitative PCR (qPCR), or real-time PCR, has been used in microbial investigations to measure the abundance and expression of taxonomic and functional gene markers. Unlike traditional PCR, which relies on end-point detection of amplified genes, qPCR uses either intercalating fluorescent dyes such as SYBR Green or fluorescent probes (TaqMan) to measure the accumulation of amplicons in real time during each cycle of the PCR. Software records the increase in amplicon concentration during the early exponential phase of amplification which enables the quantification of genes (or transcripts) when they are proportional to the starting template concentration. When qPCR is coupled with a preceding reverse transcription (RT) reaction, it can be used to quantify gene expression (RT-qPCR). qPCR is highly sensitive to starting template concentration and measures template abundance in a large dynamic range of around six orders of magnitude. Several sets of 16S and 5.8S rRNA gene primers have been designed for rapid qPCR based quantification of soil bacterial and fungal microbial communities [64]. Aparna and his co-workers investigated the quantification of 16S rDNA, *amoA* and *nifH* genes in organic and inorganic cropping soil using qPCR, which shown a clear 1.8 fold increase in both organic cropping and organic orchard soils whereas the abundance of *amoA* gene decreased by 22- and 11-folds in organic cropping and orchards [65].

5.4 Sequencing based methods

5.4.1 Clone library sequencing

Clone library sequencing involves extraction of environmental DNA followed by amplification of partial or full length of 16S rRNA (27f (5′-AGRGTTTGAT CMTGGCTCAG)) and 1492R (5′-GGTTACCTTGTTACGACTT). The amplified sequences are then ligated and cloned in a suitable vector. Further the clones containing organism-specific 16S rRNA gene fragments are purified and sequenced from each terminus. Sequences are then assembled and checked for QC. Phylum, class, order, family, subfamily, or OTU placement are determined when a clone surpasses similarity thresholds of 80, 85, 90, 92, 94, or 97%, respectively. When similarity to nearest database sequence falls below 94%, the clone is considered to represent a novel subfamily and a novel class was denoted when similarity is less than 85% [66]. A comprehensive study was done to characterize the relative abundance, diversity and composition of acidobacterial communities across a range of soil types using clone library analyses [67]. Vázquez and co-workers collected the soil and sediment samples from the coastal region in response to diesel spill, based on DGGE analysis they selected six soils and two sediment samples for identification of bacterial community structure using clone library analyses [68].

5.4.2 Amplicon sequencing

Soil DNA is extracted and then 16S/18S rDNA genes are amplified by using specific set of specific primers targeting variable regions of 16S/18S rDNA, followed by purification of fragments using magnetic beads, then adapters are ligated and the library of fragments (clones) is amplified and the samples are sequenced using NGS platform. The dataset obtained after sequencing can be compared with Ribosomal Database Project (RDP) for identification of microbial community harboring in environmental sites [69]. Using NGS, it is possible to resolve highly complex microbiota compositions with greater accuracy, as well as to link microbial community diversity with niche function. The soil DNA was extracted and the two step PCR amplification was done using domain-specific primer, 515F/806R (for prokaryotes), F1427/R1616 (for eukaryotes) or ITSF1/ITSF2 (for fungi) followed by purification and amplicon sequencing using illumina platform for identification of soil bacterial and fungal community. [70]. Schöler and colleagues has stated brief highlights regarding the crucial steps that should be considered for accurate analysis and data interpretation while opting for amplicon sequencing using marker genes [71].

5.4.3 Shotgun metagenome sequencing

Exploration of microbial universe in environmental systems through shotgun metagenome sequencing allows us to investigate deeper strata of the microbial communities and provides an unbiased view on the phylogentic and functional composition of the environmental microbial communities [21]. For sequencing of the soil or environmental DNA the steps involved are, extraction of high quality total community DNA followed by fragmentation to obtain desired length of fragments, which are further purified followed by amplification and sequencing using desired or available sequencing platform. The data set thus obtained is analyzed with the offline (MEGAN) or web- based software (MG-RAST) for visualization or comparison of the pictures of microbial world. Although shotgun metagenomic sequencing does not involve the biased amplification of 16S rRNA genes, the relative organism abundances inferred from metagenomic sequences vary significantly depending on the DNA extraction and sequencing protocol utilized [72]. Utilizing the Illumina sequencing platform, the impact of N fertilization on soil microbial communities was studied, where the field was cultivated under soybean and corn [73]. Luo and is team reported whole-genome shotgun metagenomic analysis of microbial communities of temperate grassland soils that experienced 2°C infrared heating for 10 years and observed that the heated communities showed significant shifts in composition and predicted metabolism, and these shifts were community wide as opposed to being attributable to a few taxa. Key metabolic pathways such as cellulose degradation (~13%), CO_2 production (~10%), and to nitrogen cycling (~12%), were enriched under heating, which was consistent with independent physicochemical measurements. [74].

6. Applications of metagenomics in soil environment

Soil habitats contain the greatest microbial diversity of all the other environments on earth. So far, metagenomic approaches able to scratch the surface of the genomic, metabolic and phylogenetic diversity stored in the soil metagenome. Being a most diverse and challenging environment, soil holds an unlimited resource for the discovery of novel genes, enzymes, natural products, bioactive compounds, and bioprocesses. Soil metagenomic methods, specifically isolation of soil DNA followed by construction and screening of clone libraries, enable to look at more complete picture of soil microbial communities and to better understand their interactions. This methodology is of great potential for use in the studies of soil microbial communities and their functional genes, and in the discovery of new biocatalysts for industry. The sustainable economic future of modern industrialized societies requires the development of novel molecules, enzymes, processes, products, and applications. Application of metagenomics in soil had helped many microbiologists for uncovering its huge potential by overcoming the need for culturing the microbes in pure form as well as capturing the unculturable ones. High-throughput and sensitive screening methods are employed to overcome the complexity of soil metagenome. In general, screening of soil metagenome libraries relies on metabolic activity (function-driven approach) or on nucleotide sequence (sequence driven approach) whichever is suitable and feasible. By employing one or combination of these methods, researchers can discover vast number of novel products of industrial or agricultural use.

6.1 Soil health

Soil microbes are essential for the proper functioning (i.e., health) of a soil. Yet, there is little information about how to assess the life present in soil to determine if a given soil is healthy or not. Till date, biological measures of soil health have been centered on biological functions, such as respiration or nitrogen mineralization. Soil metagenomics is a promising approach in describing the functional potential of the soil microbial community, which might yield greater insight into the health of a soil than taxonomy-based metrics. Soil health cannot be measured directly, so quality/health of soil is evaluated by indicators. Structural and functional diversity of microbes, presence/absence of important microbial players, microbial activity (respiration, DNA replication and cell division), nutrient cycling, production of cofactors and secondary metabolites, and response to biotic and abiotic stress can be used as biological indicators of soil health. Biological activity of the soil can be known by estimating dehydrogenase enzyme in the soils [75]. β-glucosidase involves in hydrolysis and biodegradation of various β-glucoside present in decomposing plant debris in soil. β-glucosidase is characteristically useful as a soil quality indicator, and may give a reflection of past biological activity, the capacity of soil to stabilize the soil organic matter, and can be used to detect management effect on soils [76]. The amount of these enzymes activity indicates the biological capacity of soil, for the enzymatic conversion of the substrate and also has an important role in the ecology of microorganisms in the ecosystem. The degradation and hydrolysis of chitin is achieved by Chitinase a key enzyme responsible for the same. Its presence in different forms in the ecosystem has demonstrated its effectiveness in the control of soil-borne diseases. Arylsulphatases are responsible for the hydrolysis of sulphate esters in the soil [77] and are secreted by bacteria into the external environment as a response to sulphur limitation. Phosphatases are good indicator of soil fertility and are believed to play critical roles in P cycles [78]. When there is a signal indicating P deficiency in the soil, acid phosphatase secretion from plant roots is increased to enhance the solubilization and remobilization of phosphate, thus influencing the ability of the plant to cope with P-stressed conditions [79].

6.2 Industrial use

Activity-based screening has the potential to detect entirely novel genes encoding new types and classes of enzymes or to identify the new bioactive compounds. In addition, it is selective for full-length genes and functional gene products. Cellulases has been isolated from various natural environments like soil, rumen, compost soil and many more using metagenomic technique by constructing the metagenomic libraries followed by screening of the biologically active clones.

Cellulases are used in animal feeds for improving the nutritional quality and digestibility, in processing of fruit juices, and in baking, while de-inking of paper is yet another emerging application [80]. Alvarez with his team isolated and characterized a novel cellulase by functional screening of a metagenomic library derived from sugarcane field land soil [65]. Lipases have been found in many species of animals and plants. The enzymes from microbial sources (such as bacteria, yeast and fungi) are currently receiving particular attention because of their actual and potential applications in industry mainly in the detergents, oils and fats, dairy and pharmaceutical industries [80]. A novel metagenomic xylanase has been isolated from compost-soil metagenome that shows alkali stability and thermostability, thus bearing a potential application in paper and pulp industry in pulp bleaching [81]. Vidya with her co-associates isolated a thermostable and calcium-dependent amylase from a soil by constructing and screening soil metagenomic library and suggested its applications in baking and de-starching [82].

6.3 Antibiotics discovery

Since the discovery of first antibiotic there has been tremendous improvement in human health. However by traditional techniques we can tap only known antibiotics identified from known organisms. However, soil metagenomics presents a greater opportunity for discovery of vast number of antibiotics yet to be discovered. Researchers have isolated new antibiotics, one of such example is Turbomycin A and B from a metagenomic library of soil microbial DNA [83]. Some of them, isolated the genes encoding enzymes required for synthesis of other antibiotics e.g., Voget and co-workers found one amidase-positive clone while general screening of a soil metagenomic library for biocatalysts [84]. Amidases are used in the biosynthesis of β-lactam antibiotics [85]. Chang and Brady reported a gene cluster *bor* from soil that encodes indolotryptoline based compounds, a small and relatively rare family of natural products that exhibit potent activity against certain tumor cell lines [86]. Through activity-based screening of *E. coli* plasmid library from four agricultural soil samples 45 clones were isolated which were showing resistant to tetracycline, chloramphenicol, kanamycin, minocycline, gentamicin, amikacin, aminoglycosides, streptomycin and rifampicin [87].

6.4 Bioremediation

Due to industrialization, dumping or burying of harmful wastes in the soil or water stream resulted in degradation of surrounding cultivable or arable lands. By tapping these affected soil environments one can discover the microbes degrading the harmful hydrocarbons. As microbes are directly involved in carbon cycles, they may play a role in breaking down the carbon present in the harmful hydrocarbons. Using a function-driven metagenomic approach, new metabolic pathways involved in the biodegradation of aromatic compounds can be discovered. Many rhizospheric microbes produce biosurfactant; these biomolecules play vital role in motility, signaling, and biofilm formation, indicating that biosurfactant governs plant–microbe interaction. In agriculture, biosurfactants can be used for plant pathogen elimination and for increasing the bioavailability of nutrient for beneficial plant associated microbes. Biosurfactants can widely be applied for improving the agricultural soil quality by soil remediation. These biomolecules can replace the harsh surfactant presently being used in million dollar pesticide industries [88]. Studies related to adaptation of microbes in toxic environments may give rise to trace new metagenomic communities useful for efficient bioremediation.

6.5 Agriculture

Soil microbes play an important role in triggering the growth, stress responses and defense in plants. Understanding the relationship between the soil microbiota and plants using soil metagenomics will be very helpful in designing the crop systems. Metagenomics study of the soils supplemented with organic manures obtained from various farm animals would be very much helpful in formulating the fertilization strategies and reducing the dependence on inorganic fertilizers. For sustainable agricultural production, beneficial microbes of agricultural importance can serve as a crucial alternative. Metagenomics help in the prediction of microbial community structure and, therefore, can tackle and address fundamental scientific questions related to agriculturally important microorganisms. This approach has been successfully explored for the assessment of the diazotrophs belonging to the rhizosphere of native red kidney beans (RKB) of the Western Indian Himalaya by targeting *nifH* [89].

7. Conclusion

Soil being a complex environment, understanding of the interactions within microbial communities and soil environments is very much necessary for designing the strategies for sustainable agriculture, bioremediation, and human welfare. Soil metagenomic data will reveal potential activities present in microbial communities which can be harnessed for future good use. The huge diversity of soil microorganisms, together with the heterogeneity of the soil environment, hinders analyses of microbial diversity, structure and the linking functional processes. One of the major challenges for soil metagenomics is to develop methods to capture the heterogeneity and dynamics of complex soil microbial communities, both over time and spatially. However, new methods will greatly increase the number of samples that can be analyzed in the future. All the methods used for the investigating microbial diversity and activity contain inherent biases and it is necessary to understand the underlying mechanisms in order to be aware of the drawbacks and limitations, and to appreciate the strengths and weaknesses of each approach [9]. Nevertheless, these methods are starting to dissect the soil microbial biomass and the soil metagenome, and will, in the future, enable a greatly improved understanding of microbial community dynamics and interactions relevant to soil functions.

The TerraGenome International Soil Metagenome Sequencing Consortium has been dedicated towards soil metagenomics and helps in co-coordinating the global researchers for the discovery of the soil hidden treasures (http://www.terragenome.org/).

Author details

Shrinivas N. Sabale, Padmaja P. Suryawanshi and Krishnaraj P.U.*
University of Agricultural Sciences, Dharwad, Karnataka, India

*Address all correspondence to: krishnarajpu@gmail.com

References

[1] Aktar W, Sengupta D, Chowdhury A. Impact of pesticides use in agriculture: Their benefits and hazards. Interdisciplinary Toxicology. 2009;**2**(1):1-2. DOI: 10.2478/v10102-009-0001-7

[2] Christopher J. Living Soils: The Role of Microorganisms in Soil Health. Strategic Analysis Paper. Nedlands: Future Directions International; 2017

[3] Handelsman J, Rondon MR, Brady SF, Clardy J, Goodman RM. Molecular biological access to the chemistry of unknown soil microbes: A new frontier for natural products. Chemistry & Biology. 1998;**5**(10):R245-R249. DOI: 10.1016/S1074-5521(98)90108-9

[4] Berendsen RL, Pieterse CM, Bakker PA. The rhizosphere microbiome and plant health. Trends in Plant Science. 2012;**17**(8):478-486. DOI: 10.1016/j.tplants.2012.04.001

[5] Daniel R. The soil metagenome—A rich resource for the discovery of novel natural products. Current Opinion in Biotechnology. 2004;**15**(3):199-204. DOI: 10.1016/j.copbio.2004.04.005

[6] Hernandez-Leon R, Martínez-Trujillo M, Valencia-Cantero E, Santoyo G. Construction and characterization of a metagenomic DNA library from the rhizosphere of wheat (*Triticum aestivum*). Phyton (Buenos Aires). 2012;**81**:133-137

[7] Bashir Y, Pradeep Singh S, Kumar Konwar B. Metagenomics: An application based perspective. Chinese Journal of Biology. 2014:1-7. DOI: 10.1155/2014/146030

[8] Doornbos RF, van Loon LC, Bakker PA. Impact of root exudates and plant defense signaling on bacterial communities in the rhizosphere. A review. Agronomy for Sustainable Development. 2012;**32**(1):227-243. DOI: 10.1007/s13593-011-0028-y

[9] Hirsch PR, Mauchline TH, Clark IM. Culture-independent molecular techniques for soil microbial ecology. Soil Biology and Biochemistry. 2010;**42**(6):878-887. DOI: 10.1016/j.soilbio.2010.02.019

[10] National Research Council. The New Science of Metagenomics: Revealing the Secrets of our Microbial Planet. Washington, DC: National Academies Press; 2007

[11] Hartmann A, Rothballer M, Schmid M. Lorenz Hiltner, a pioneer in rhizosphere microbial ecology and soil bacteriology research. Plant and Soil. 2008;**312**(1-2):7-14. DOI: 10.1007/s11104-007-9514-z

[12] Kirk JL, Beaudette LA, Hart M, Moutoglis P, Klironomos JN, Lee H, et al. Methods of studying soil microbial diversity. Journal of Microbiological Methods. 2004;**58**(2):169-188. DOI: 10.1016/j.mimet.2004.04.006

[13] Wu Z, Lin W, Li B, Wu L, Fang C, Zhang Z. Terminal restriction fragment length polymorphism analysis of soil bacterial communities under different vegetation types in subtropical area. PLoS One. 2015;**10**(6):e0129397. DOI: 10.1371/journal.pone.0129397

[14] Garland JL, Mills AL. Classification and characterization of heterotrophic microbial communities on the basis of patterns of community-level sole-carbon-source utilization. Applied and Environmental Microbiology. 1991;**57**(8):2351-2359

[15] Stefanowicz A. The biolog plates technique as a tool in ecological studies of microbial communities. Polish Journal of Environmental Studies. 2006;**15**(5):669-676

[16] Al-Dhabaan FA, Bakhali AH. Analysis of the bacterial strains using Biolog plates in the contaminated soil from Riyadh community. Saudi Journal of Biological Sciences. 2017;**24**(4):901-906. DOI: 10.1016/j.sjbs.2016.01.043

[17] Joergensen RG, Wichern F. Quantitative assessment of the fungal contribution to microbial tissue in soil. Soil Biology and Biochemistry. 2008;**40**(12):2977-2991. DOI: 10.1016/j.soilbio.2008.08.017

[18] Docherty KM, Borton HM, Espinosa N, Gebhardt M, Gil-Loaiza J, Gutknecht JL, et al. Key edaphic properties largely explain temporal and geographic variation in soil microbial communities across four biomes. PLoS One. 2015;**10**(11):e0135352. DOI: 10.1371/journal.pone.013535

[19] Chodak M, Gołębiewski M, Morawska-Płoskonka J, Kuduk K, Niklińska M. Diversity of microorganisms from forest soils differently polluted with heavy metals. Applied Soil Ecology. 2013;**64**:7-14. DOI: 10.1016/j.apsoil.2012.11.004

[20] Miura T, Makoto K, Niwa S, Kaneko N, Sakamoto K. Comparison of fatty acid methyl ester methods for characterization of microbial communities in forest and arable soil: Phospholipid fraction (PLFA) versus total ester linked fatty acids (EL-FAME). Pedobiologia. 2017;**63**:14-18. DOI: 10.1016/j.pedobi.2017.04.002

[21] Rastogi G, Sani RK. Molecular techniques to assess microbial community structure, function, and dynamics in the environment. In: Ahmad et al., editors. Microbes and Microbial Technology: Agricultural and Environmental Applications. New York: Springer Science+Business Media, LLC; 2011. DOI: 10.1007/978-1-4419-7931-5_2

[22] Sharmili AS, Ramasamy P. Fatty acid methyl ester (FAME) analysis of moderately thermophilic bacteria isolated from the Coramandel Coast, Chennai, Tamilnadu. European Journal of Experimental Biology. 2016;**6**(5):1-7

[23] Torsvik V, Daae FL, Sandaa RA, Øvreås L. Novel techniques for analysing microbial diversity in natural and perturbed environments. Journal of Biotechnology. 1998;**64**(1):53-62. DOI: 016/S0168-1656(98)00103-5

[24] Fakruddin M, Mannan K. Methods for analyzing diversity of microbial communities in natural environments. Ceylon Journal of Science (Biological Sciences). 2013;**42**(1):19-33. DOI: 10.4038/cjsbs.v42i1.5896

[25] Nüsslein K, Tiedje JM. Soil bacterial community shift correlated with change from forest to pasture vegetation in a tropical soil. Applied and Environmental Microbiology. 1999;**65**(8):3622-3626

[26] Holben WE. GC Fractionation allows comparative total microbial community analysis, enhances diversity assessment, and facilitates detection of minority populations of bacteria. In: Handbook of Molecular Microbial Ecology. I: Metagenomics and Complementary Approaches. New York: John Wiley & Sons, Inc.; 2011. pp. 183-196

[27] Greene EA, Voordouw G. Analysis of environmental microbial communities by reverse sample genome probing. Journal of Microbiological Methods. 2003;**53**(2):211-219. DOI: 10.1016/S0167-7012(03)00024-1

[28] Williams JG, Kubelik AR, Livak KJ, Rafalski JA, Tingey SV. DNA polymorphisms amplified by arbitrary primers are useful as genetic markers. Nucleic Acids Research. 1990;**18**(22):6531-6535

[29] Sharma G, Verma HN, Sharma R. RAPD analysis to study metagenome diversity in soil microbial community

of arid zone plants. Proceedings of the National Academy of Sciences, India Section B: Biological Sciences. 2013;**83**(2):135-139. DOI: 10.1007/s40011-012-0041-1

[30] Tilwari A, Chouhan D, Sharma R. Random amplified polymorphic DNA (RAPD) analysis of microbial community diversity in soil affected by industrial pollutants: Reference to Mandideep Industrial Area. African Journal of Microbiology Research. 2013;**7**(30):3933-3942. DOI: 10.5897/AJMR12.1720

[31] Smit E, Leeflang P, Wernars K. Detection of shifts in microbial community structure and diversity in soil caused by copper contamination using amplified ribosomal DNA restriction analysis. FEMS Microbiology Ecology. 1997;**23**(3):249-261. DOI: 10.1111/j.1574-6941.1997.tb00407.x

[32] Sklarz MY, Angel R, Gillor O, Soares MI. Evaluating amplified rDNA restriction analysis assay for identification of bacterial communities. Antonie Van Leeuwenhoek. 2009;**96**(4):659. DOI: 10.1007/s10482-009-9380-1

[33] Liu WT, Marsh TL, Cheng H, Forney LJ. Characterization of microbial diversity by determining terminal restriction fragment length polymorphisms of genes encoding 16S rRNA. Applied and Environmental Microbiology. 1997;**63**(11):4516-4522

[34] Leckie SE. Methods of microbial community profiling and their application to forest soils. Forest Ecology and Management. 2005;**220**(1-3):88-106. DOI: 10.1016/j.foreco.2005.08.007

[35] Singh BK, Nazaries L, Munro S, Anderson IC, Campbell CD. Use of multiplex terminal restriction fragment length polymorphism for rapid and simultaneous analysis of different components of the soil microbial community. Applied and Environmental Microbiology. 2006;**72**(11):7278-7285. DOI: 10.1128/AEM.00510-06

[36] Joo S, Lee SR, Park S. Monitoring of phytoplankton community structure using terminal restriction fragment length polymorphism (T-RFLP). Journal of Microbiological Methods. 2010;**81**(1):61-68. DOI: 10.1016/j.mimet.2010.01.025

[37] Borneman J, Triplett EW. Molecular microbial diversity in soils from eastern Amazonia: Evidence for unusual microorganisms and microbial population shifts associated with deforestation. Applied and Environmental Microbiology. 1997;**63**:2647-2653

[38] Fisher MM, Triplett EW. Automated approach for ribosomal intergenic spacer analysis of microbial diversity and its application to freshwater bacterial communities. Applied and Environmental Microbiology. 1999;**65**(10):4630-4636

[39] Delmont TO, Robe P, Cecillon S, Clark IM, Constancias F, Simonet P, et al. Accessing the soil metagenome for studies of microbial diversity. Applied and Environmental Microbiology. 2011;**77**(4):1315-1324. DOI: 10.1128/AEM.01526-10

[40] Muyzer G, De Waal EC, Uitterlinden AG. Profiling of complex microbial populations by denaturing gradient gel electrophoresis analysis of polymerase chain reaction-amplified genes coding for 16S rRNA. Applied and Environmental Microbiology. 1993;**59**(3):695-700

[41] Wang Y, Qian PY. Conservative fragments in bacterial 16S rRNA genes and primer design for 16S ribosomal DNA amplicons in metagenomic studies. PLoS One. 2009;**4**(10):e7401. DOI: 10.1371/journal.pone.0007401

[42] Carvajal AM, Vargas RA, Alfaro M. Abundance of denitrifying genes and microbial community structure in volcanic soils. Journal of Soil Science and Plant Nutrition. 2016;**16**(3):677-688. DOI: 10.4067/S0718-95162016005000049

[43] Marzorati M, Wittebolle L, Boon N, Daffonchio D, Verstraete W. How to get more out of molecular fingerprints: Practical tools for microbial ecology. Environmental Microbiology. 2008;**10**(6):1571-1581. DOI: 10.1111/j.1462-2920.2008.01572.x

[44] Luo P, Han X, Wang Y, Han M, Shi H, Liu N, et al. Influence of long-term fertilization on soil microbial biomass, dehydrogenase activity, and bacterial and fungal community structure in a brown soil of Northeast China. Annals of Microbiology. 2015;**65**(1):533-542. DOI: 10.1007/s13213-014-0889-9

[45] Ding GC, Piceno YM, Heuer H, Weinert N, Dohrmann AB, Carrillo A, et al. Changes of soil bacterial diversity as a consequence of agricultural land use in a semi-arid ecosystem. PLoS One. 2013;**8**(3):e59497. DOI: 10.1371/journal.pone.0059497

[46] Jost L. The relation between evenness and diversity. Diversity. 2010;**2**(2):207-232. DOI: 10.3390/d2020207

[47] Peet RK. The measurement of species diversity. In: Section of Ecology and Systematic. Vol. 5. Ithaca, New York: Langmuir Laboratory, Cornell University; 1974. pp. 285-307

[48] Shannon CE. A mathematical theory of communication. Bell System Technical Journal. 1948;**27**(3):379-423

[49] Magurran AE. Ecological Diversity and its Measurement. Princeton, New Jersey: Princeton University Press; 1988

[50] Chao A, Chazdon RL, Colwell RK, Shen TJ. A new statistical approach for assessing similarity of species composition with incidence and abundance data. Ecology Letters. 2005;**8**(2):148-159. DOI: 10.1111/j.1461-0248.2004.00707.x

[51] Seghers D, Top EM, Reheul D, Bulcke R, Boeckx P, Verstraete W, et al. Long-term effects of mineral versus organic fertilizers on activity and structure of the methanotrophic community in agricultural soils. Environmental Microbiology. 2003;**5**(10):867-877. DOI: 10.1046/j.1462-2920.2003.00477.x

[52] Possemiers S, Verthé K, Uyttendaele S, Verstraete W. PCR-DGGE-based quantification of stability of the microbial community in a simulator of the human intestinal microbial ecosystem. FEMS Microbiology Ecology. 2004;**49**(3):495-507. DOI: 10.1016/j.femsec.2004.05.002

[53] Lozada M, Figuerola EL, Itria RF, Erijman L. Replicability of dominant bacterial populations after long-term surfactant-enrichment in lab-scale activated sludge. Environmental Microbiology. 2006;**8**(4):625-638. DOI: 10.1111/j.1462-2920.2005.00939.x

[54] LaPara TM, Nakatsu CH, Pantea LM, Alleman JE. Stability of the bacterial communities supported by a seven-stage biological process treating pharmaceutical wastewater as revealed by PCR-DGGE. Water Research. 2002;**36**(3):638-646. DOI: 10.1016/S0043-1354(01)00277-9

[55] Theron J, Cloete TE. Molecular techniques for determining microbial diversity and community structure in natural environments. Critical Reviews in Microbiology. 2000;**26**(1):37-57. DOI: 10.1080/10408410091154174

[56] Gentry TJ, Wickham GS, Schadt CW, He Z, Zhou J. Microarray

applications in microbial ecology research. Microbial Ecology. 2006;**52**(2):159-175. DOI: 10.1007/s00248-006-9072-6

[57] Zhou J, He Z, Yang Y, Deng Y, Tringe SG, Alvarez-Cohen L. High-throughput metagenomic technologies for complex microbial community analysis: open and closed formats. MBio;**6**:e02288-14. DOI: 10.1128/mBio.02288-14

[58] Zengler K, Walcher M, Clark G, Haller I, Toledo G, Holland T, et al. High-throughput cultivation of microorganisms using microcapsules. Methods in Enzymology. 2005;**397**:124-130. DOI: 10.1016/S0076-6879(05)97007-9

[59] Schmalenberger A, Tebbe CC. Profiling the diversity of microbial communities with single-strand conformation polymorphism (SSCP). In: Environmental Microbiology. Totowa, NJ: Humana Press; 2014. pp. 71-83. DOI: 10.1007/978-1-62703-712-9_6

[60] Smalla K, Oros-Sichler M, Milling A, Heuer H, Baumgarte S, Becker R, et al. Bacterial diversity of soils assessed by DGGE, T-RFLP and SSCP fingerprints of PCR-amplified 16S rRNA gene fragments: Do the different methods provide similar results? Journal of Microbiological Methods. 2007;**69**(3):470-479. DOI: 10.1016/j.mimet.2007.02.014

[61] Hjelmsø MH, Hansen LH, Bælum J, Feld L, Holben WE, Jacobsen CS. High-resolution melt analysis for rapid comparison of bacterial community compositions. Applied and Environmental Microbiology. 2014;**80**(12):3568-3575. DOI: 10.1128/AEM.03923-13

[62] Arabet D, Tempel S, Fons M, Denis Y, Jourlin-Castelli C, Armitano J, et al. Effects of a sulfonylurea herbicide on the soil bacterial community. Environmental Science and Pollution Research. 2014;**21**(8):5619-5627. DOI: 10.1007/s11356-014-2512-9

[63] Yergeau E, Bezemer TM, Hedlund K, Mortimer SR, Kowalchuk GA, Van Der Putten WH. Influences of space, soil, nematodes and plants on microbial community composition of chalk grassland soils. Environmental Microbiology. 2010;**12**(8):2096-2106. DOI: 10.1111/j.1462-2920.2009.02053.x

[64] Fierer N, Jackson JA, Vilgalys R, Jackson RB. Assessment of soil microbial community structure by use of taxon-specific quantitative PCR assays. Applied and Environmental Microbiology. 2005;**71**(7):4117-4120. DOI: 10.1128/AEM.71.7.4117-4120.2005

[65] Aparna K, Pasha MA, Rao DLN, Krishnaraj PU. Organic amendments as ecosystem engineers: Microbial, biochemicaland genomic evidence of soil health improvement in a tropical aridzone field site. Ecological Engineering. 2014;**71**:268-277. DOI: 10.1016/j.ecoleng.2014.07.016

[66] DeSantis TZ, Brodie EL, Moberg JP, Zubieta IX, Piceno YM, Andersen GL. High-density universal 16S rRNA microarray analysis reveals broader diversity than typical clone library when sampling the environment. Microbial Ecology. 2007;**53**(3):371-383. DOI: 10.1007/s00248-006-9134-9

[67] Jones RT, Robeson MS, Lauber CL, Hamady M, Knight R, Fierer N. A comprehensive survey of soil acidobacterial diversity using pyrosequencing and clone library analyses. The ISME Journal. 2009;**3**(4):442-453. DOI: 10.1038/ismej.2008.127

[68] Vázquez S, Monien P, Minetti RP, Jürgens J, Curtosi A, Primitz JV, et al.

Bacterial communities and chemical parameters in soils and coastal sediments in response to diesel spills at Carlini Station, Antarctica. Science of the Total Environment. 2017;**605**:26-37. DOI: 10.1016/j.scitotenv.2017.06.129

[69] Rosselli R, Romoli O, Vitulo N, Vezzi A, Campanaro S, De Pascale F, et al. Direct 16S rRNA-seq from bacterial communities: A PCR-independent approach to simultaneously assess microbial diversity and functional activity potential of each taxon. Scientific Reports. 2016;**6**:32165. DOI: 10.1038/srep32165

[70] Tkacz A, Hortala M, Poole PS. Absolute quantitation of microbiota abundance in environmental samples. Microbiome. 2018;**6**(1):110. DOI: 10.1186/s40168-018-0491-7

[71] Schöler A, Jacquiod S, Vestergaard G, Schulz S, Schloter M. Analysis of soil microbial communities based on amplicon sequencing of marker genes. Biology and Fertility of Soils. 2017;**53**(5):485-489. DOI: 10.1007/s00374-017-1205-1

[72] Morgan JL, Darling AE, Eisen JA. Metagenomic sequencing of an in vitro-simulated microbial community. PLoS One. 2010;**5**(4):e10209. DOI: 10.1371/journal.pone.0010209

[73] Orellana LH, Chee-Sanford JC, Sanford RA, Löffler FE, Konstantinidis KT. Year-round shotgun metagenomes reveal stable microbial communities in agricultural soils and novel ammonia oxidizers responding to fertilization. Applied and Environmental Microbiology. 2018;**84**(2):e01646-e01617. DOI: 10.1128/AEM.01646 17

[74] Luo C, Rodriguez-R LM, Johnston ER, Wu L, Cheng L, Xue K, et al. Soil microbial community responses to a decade of warming as revealed by comparative metagenomics. Applied and Environmental Microbiology. 2014;**80**(5):1777-1786. DOI: 10.1128/AEM.03712-13

[75] Burns RG. Enzyme activity in soil: Some theoretical and practical considerations. In: Soil Enzymes. London: Academic Press; 1978. pp. 295-340

[76] Ndiaye EL, Sandeno JM, McGrath D, Dick RP. Integrative biological indicators for detecting change in soil quality. American Journal of Alternative Agriculture. 2000;**15**(1):26-36. DOI: 10.1017/S0889189300008432

[77] Kertesz MA, Mirleau P. The role of soil microbes in plant sulphur nutrition. Journal of Experimental Botany. 2004;**55**(404):1939-1945. DOI: 10.1093/jxb/erh176

[78] Speir TW, Ross DJ. Soil phosphatase and sulphatase. In: Burns RG, editor. Soil Enzymes. London, UK: Academic; 1978. pp. 197-250

[79] Versaw WK, Harrison MJ. A chloroplast phosphate transporter, PHT2; 1, influences allocation of phosphate within the plant and phosphate-starvation responses. The Plant Cell. 2002;**14**(8):1751-1766. DOI: 10.1105/tpc.002220

[80] Nazir A. Review on metagenomics and its applications. Imperial Journal of Interdisciplinary Research. 2016;**2**:10

[81] Verma D, Kawarabayasi Y, Miyazaki K, Satyanarayana T. Cloning, expression and characteristics of a novel alkalistable and thermostable xylanase encoding gene (Mxyl) retrieved from compost-soil metagenome. PLoS One. 2013;**8**(1):e52459. DOI: 10.1371/journal.pone.0052459

[82] Vidya J, Swaroop S, Singh SK, Alex D, Sukumaran RK, Pandey A. Isolation and characterization of a novel α-amylase from a metagenomic library

of Western Ghats of Kerala, India. Biologia. 2011;**66**(6):939. DOI: 10.2478/s11756-011-0126-y

[83] Gillespie DE, Brady SF, Bettermann AD, Cianciotto NP, Liles MR, Rondon MR, et al. Isolation of antibiotics turbomycin a and B from a metagenomic library of soil microbial DNA. Applied and Environmental Microbiology. 2002;**68**(9):4301-4306. DOI: 10.1128/AEM.68.9.4301-4306

[84] Voget S, Leggewie C, Uesbeck A, Raasch C, Jaeger KE, Streit WR. Prospecting for novel biocatalysts in a soil metagenome. Applied and Environmental Microbiology. 2003;**69**(10):6235-6242. DOI: 10.1128/AEM.69.10.6235-6242.2003

[85] Ghazanfar S, Azim A, Ghazanfar MA, Anjum MI, Begum I. Metagenomics and its application in soil microbial community studies: Biotechnological prospects. Journal of Animal & Plant Sciences. 2010;**6**(2):611-622

[86] Chang F, Brady SF. Discovery of indolotryptoline antiproliferative agents by homology-guided metagenomic screening. Proceedings of the National Academy of Sciences of the United States of America. 2013;**110**(7):2478-2483. DOI: 10.1073/pnas.1218073110

[87] Su JQ, Wei B, Xu CY, Qiao M, Zhu YG. Functional metagenomic characterization of antibiotic resistance genes in agricultural soils from China. Environment International. 2014;**65**:9-15. DOI: 10.1016/j.envint.2013.12.010

[88] Sachdev DP, Cameotra SS. Biosurfactants in agriculture. Applied Microbiology and Biotechnology. 2013;**97**(3):1005-1016. DOI: 10.1007/s00253-012-4641-8

[89] Suyal DC, Yadav A, Shouche Y, Goel R. Diversified diazotrophs associated with the rhizosphere of Western Indian Himalayan native red kidney beans (*Phaseolus vulgaris* L.). 3 Biotech. 2015;**5**:433-441. DOI: 10.1007/s13205-014-0238-5

Section 2

Metagenomic Analysis

Chapter 3

High-Throughput Sequencing and Metagenomic Data Analysis

Ahmed Shuikan, Sulaiman Ali Alharbi,
Dalal Hussien M. Alkhalifah and Wael N. Hozzein

Abstract

Metagenomic approaches are a growing branch of science and have many applications in different fields. Metagenomics seems to be the ideal culture-independent technique for unraveling the biodiversity of soils and to study how this biodiversity is affected with continuously changing conditions. In addition, its application in clinical and diagnostic approaches was reported. The emergence of several next-generation sequencing (NGS) strategies enriched the metagenomics. The combination between NGS and metagenomic approaches helped the investigators resolve several issues regarding the microbial diversity and the functions and relationships among different microbial flora. A number of NGS approaches were developed including Roche/454 pyrosequencing, Illumina/Solexa sequencing, and Applied Biosystems/SOLiD sequencing. In this chapter, different NGS platforms are discussed in terms of principle, advantages, and limitations. In addition, third-generation sequencing technologies are also addressed.

Keywords: high throughput, metagenomics workflow, sequencing approaches, metagenomic data analysis

1. Introduction

The development of next-generation sequencing (NGS) techniques provides high-throughput sequence analysis with the ability to simultaneously and independently sequence billions of DNA molecules. The combination between such technologies and metagenomic approaches helped the investigators study the microbial diversity and understand the functions and relationships among different microbial flora [1]. The use of metagenomic NGS by microbiologists overcomes several limitations and secured the unbiased methods to study the microbial flora in any given environment [2]. Thus, the dynamic of complex communities particularly those with non-cultivable microorganisms can be resolved [3, 4]. In addition, metagenomic NGS found its way in the field of clinical and diagnostic approaches [5, 6]. In the clinical field, NGS was used to inform the real-time incidence and prevention response to human parainfluenza 3 virus infections [7] and for cerebrospinal fluid diagnostics [8].

Several NGS platforms were developed since 2006 with numerous applications in genetic and biological research fields. Of these platforms, the most commonly used include Roche/454 pyrosequencing, Illumina/Solexa sequencing, and Applied Biosystems/SOLiD sequencing. The principle of all these NGS depends on the detection of luminescent signals released by the base incorporation during the sequencing process [4]. They also share the same workflow which in the order, DNA extraction, library construction, DNA template preparation, and automated sequence analysis [9]. In this chapter, different NGS platforms are discussed in terms of principle, advantages, and limitations. In addition, third-generation sequencing technologies are also addressed.

2. Workflow of metagenomics

2.1 The sampling process and library construction for metagenomic analysis

Metagenomic analysis is a sophisticated process and involves several steps. Of these steps, the sampling process is very crucial for the downstream applications. Sample collection, preparation, and storage should be handled carefully to prevent lysis and decomposition of the sample compositions. Multiple freezing–thawing cycles may cause changes in the microbial community profile under investigation [10]. As well, a suitable DNA extraction protocol should be adopted to cope with the different chemical and physical characteristics of each sample. For instance, soils contain many substances that are co-extracted with the genomic DNA and may have inhibitory effects on the downstream experiments. Examples include humic and fulvic acids [11]. Therefore, optimization and comparison between different extraction methods are usually required for each type of samples [12, 13, 14, 15].

The extracted DNA is used to construct the DNA library. This is usually achieved by connecting specific adaptors to one or both ends of the DNA fragments [16]. The reason for utilizing DNA adaptor is to deal with the pool of samples and then connect them to its original sample. Handling DNA at this stage should be careful to avoid chemical, physical, or enzymatic damage of DNA molecules [17]. The construction of a DNA library is usually achieved through two approaches. The first one is called meta-pair where the library is characterized by long fragment insert. The second approach is called paired-end libraries with short fragment insert. In both approaches, the DNA is fragmented into different fragment sizes that would allow for their cloning. The DNA fragments obtained from such processes are cloned into the proper cloning vector. The size of the resulting fragments determines the suitable vector for the cloning process. The small DNA fragments are usually cloned into plasmid vectors, whereas fragments up to 40 kbp are cloned into cosmid or fosmid vectors. Bacterial artificial chromosome (BAC) vectors are usually used to clone inserts with sizes that exceed 40 Kbp [18]. Finally, the free adaptor, dimers of the adaptor, and any other artifacts must be removed to avoid noisy sequencing data [17].

2.2 Sequencing approaches

During the 1970s, the first-generation sequencing techniques, chain termination [19], and chemical sequencing approaches [20] were developed. In contrast to the chemical sequencing approach, the Sanger sequencing method ultimately prevailed and found immense applications due to its simplicity and is more amenable to being scaled up [21]. Simply, the basis of Sanger sequencing depends on the incubation of a specific primer and the template DNA in the presence of DNA polymerase. The reaction is accomplished by the addition of a mixture of deoxyribonucleotide triphosphates and dNTPs' dideoxyribonucleotide triphosphates for chain termination, one of which was labeled with phosphorus-32. The resulting pool of DNA amplicons will be with the same 5′ residue and different dNTP residues at the 3′ end (**Figure 1**). This pool of DNA fragments is then fractionated by denaturing polyacrylamide gel electrophoresis giving a band pattern. In this way, DNA decoding can be achieved

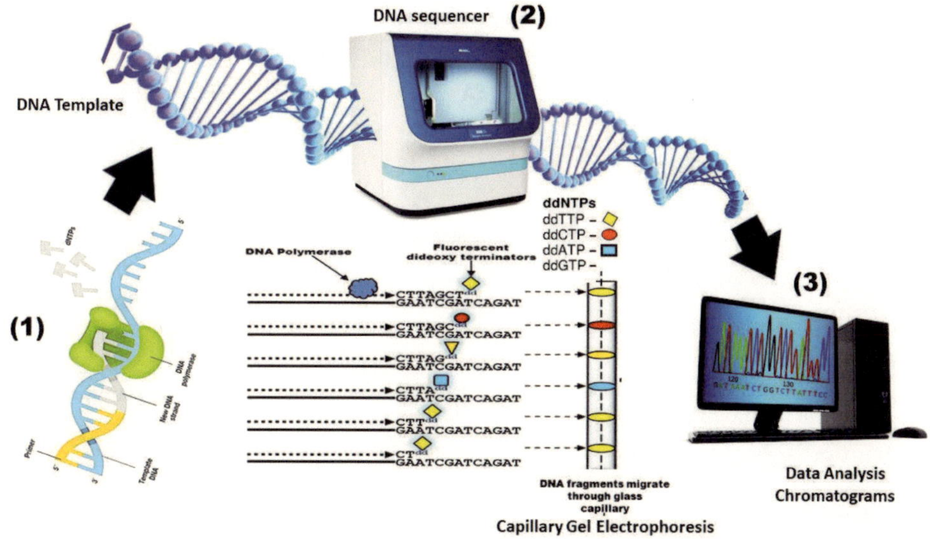

Figure 1.
Sanger DNA sequencing. (1) The gene to be decoded is amplified by PCR. (2) The sequencing process is performed by the addition of modified 2',3'-dideoxynucleotide (ddNTPs) to the nascent chain. The modified nucleotides act by terminating the chain extension, and the resulting DNA fragments of different sizes are eluted by capillary gel electrophoresis. (3) Chromatograms are then analyzed to obtain the DNA sequences.

by the use of nucleotide analogs and other nucleotides in separate incubations and concomitant electrophoretic analysis [22]. Currently, the use of fluorescent dNTPs associated with the capillary electrophoresis provides full automation of the Sanger approach. This modification allows retrieving up to 96 sequences per run with an average 800–1000 bp size of DNA fragments [21, 23, 24]. Although the Sanger sequencing was the mainstay of the original human genome project, this approach still has some limitations. These limitations include high cost and low throughput, and it is inadequate for studying unculturable organisms in complex environments [25].

2.2.1 Next-generation sequencing (NGS)

Due to the limitations of Sanger sequencing technique, next-generation sequencing emerged in 2005 [26]. Indeed, next-generation sequencing has made it possible to study and identify organisms directly from their habitats without prior preparations [27]. Compared to the first-generation sequencing, NGS can generate several hundred thousand to millions of sequencing reads in parallel. As well, sequencing can be generated without some conventional steps such as vector-based cloning procedure and hence reduces the chance of DNA contamination from other organisms [28]. Therefore, several next-generation sequencing platforms have been introduced including Roche 454, Illumina®, Applied Biosystems SOLiD sequencer, and Ion Torrent. All next-generation sequencing or real-time sequencing (Roche 454, Illumina®, and AB SOLiD) utilized optical sensors that detect luminescent signal, which are produced during incorporation of bases in the sequence. The principles and characteristics of NSG, SGS, and TGS are summarized in **Table 1** [21]. In the subsequent sections, the features and limitations of each of the NGS techniques are discussed.

2.2.1.1 Roche 454 genome sequence

Roche/454 pyrosequencing is the first NGS technology that launched and became commercially available in 2005. It uses real-time sequencing-by-synthesis

	First generation	Second generation	Third generation
Fundamental technology	Size separation of specifically end-labeled DNA fragments, produced by SBS or degradation	Wash-and-scan SBS	SBS, by degradation, or direct physical inspection of the DNA molecule
Resolution	Averaged across many copies of the DNA molecule being sequenced	Averaged across many copies of the DNA molecule being sequenced	Single-molecule resolution
Current raw read accuracy	High	High	Moderate
Current read length	Moderate (800–1000 bp)	Short, generally much shorter than Sanger sequencing	Long, 1000 bp, and longer in commercial systems
Current throughput	Low	High	Moderate
Current cost	High cost per base	Low cost per base	Low-to-moderate cost per base
RNA sequencing method	cDNA sequencing	cDNA sequencing	Direct RNA sequencing and cDNA sequencing
Time from start of sequencing reaction to result	Hours	Days	Hours
Sample preparation	Moderately complex, PCR amplification not required	Complex, PCR amplification required	Ranges from complex to very simple depending on technology
Data analysis	Routine	Complex because of large data volumes and because short reads complicate assembly and alignment algorithms	Complex because of large data volumes and because technologies yield new types of information and new signal processing challenges
Primary results	Base calls with quality values	Base calls with quality values	Base calls with quality values, potentially other base information such as kinetics

Table 1.
The features and principles of first-generation sequencing, SGS, and TGS.

(SBS) pyrosequencing technology, and it depends on the detection of pyrophosphate (PPi) molecule that is initiated from the incorporation of a nucleotide in the DNA polymerase (**Figure 2**) [29]. Briefly, the 454 pyrosequencing technology is proceeding as follows: (i) the library fragments are connected to beads that carry oligonucleotides complementary to adapter sequence ligated at the ends, (ii) amplifying the library fragments by emulsion PCR resulting in DNA beads that carry millions of copies of DNA fragments on their surface, and (iii) the amplified beads are inserted into picotiter plate (PTP) that consists of millions of wells. Each well can hold only one amplified bead and contains diluted pyrosequence enzyme beads, DNA amplified beads, PPiase beads, and pyrosequence beads. Finally, the light emission from PTP is recorded by a CCD camera and is translated to nucleotide sequences [29]. In comparison with other NGS platform, 454 pyrosequencing has

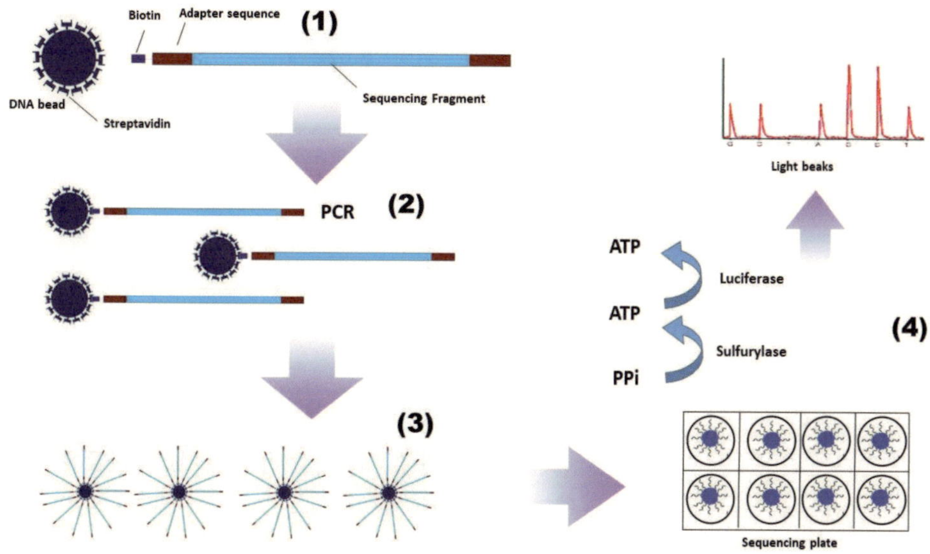

Figure 2.
Pyrosequencing technique. (1) Beads coated with either streptavidin or complementary oligonucleotides complementary to adapter sequences attached to the ends of the fragment to be sequenced. This allows the binding of sequencing fragments to the beads. (2) The fragments to be sequenced are amplified through emulsion PCR. (3) Loaded beads are transferred into the sequencing plate with millions of wells. (4) By the addition of a nucleotide to the nascent chain that is connected to the beads by DNA polymerase, the ATP sulfurylase enzyme converts released pyrophosphate to ATP with the emission of light that is detected by a CCD camera and is translated to nucleotide sequences.

the longest reading (up to 1000–1200 bp). On the other hand, 454 pyrosequencing has the highest cost per base and the lowest output [30].

2.2.1.2 Illumina sequencing (Solexa genome analyzer)

Illumina, formerly known as Solexa, has been introduced commercially in 2007. Illumina technology utilizes bridge PCR amplification coupled with SBS in the flow cell (**Figure 3**). Simply, the principle of Illumina sequencing is that the DNA fragments with barcoding primer (adaptor) are attached to the flow cell. The sequencing reaction is performed in the flow cell by adding labeled nucleotides. When the nucleotide is incorporated, a luminescent signal is generated and then recorded by optical sensors. After that, the fluorescent molecules are removed and the next labeled nucleotide incorporated. However, the DNA fragment can be sequenced on one side that is called single-end (SE) or from both sides known as paired-end (PE). Nowadays, the most common sequencing used is PE due to the ability to generate two reads for one DNA fragment which is useful in order to determine the distance between two ends of the DNA fragment [31]. In fact, due to its low cost per base and high yield, Illumina becomes the most widely used and popular NGS platform. The output of Illumina sequencing is the highest among all NGS, making it suitable for multiplexing hundreds of samples at the same time [32].

2.2.1.3 Applied biosystems (AB) SOLiD sequencer

AB SOLiD refers to sequencing by oligonucleotide ligation and detection. It has been developed by Applied Biosystems (Life Technology) and became commercially available in 2007. The AB SOLiD sequencing approach differs from the other

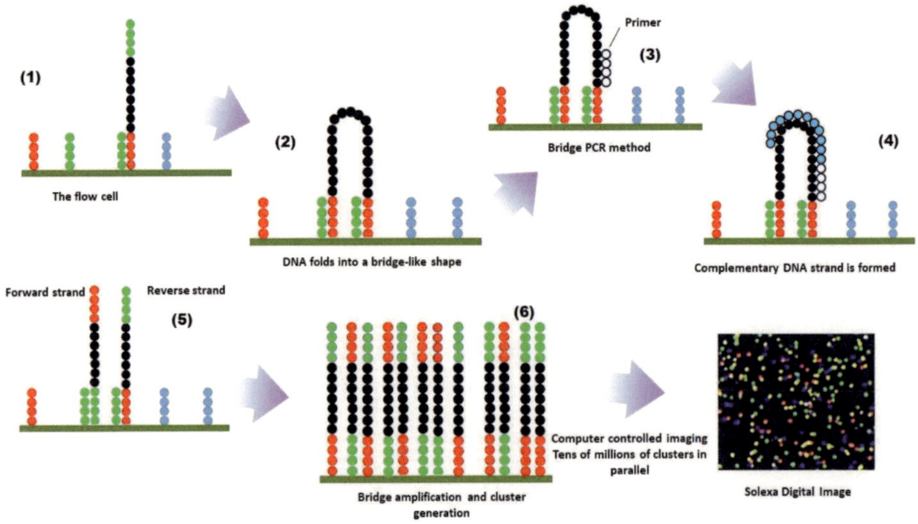

Figure 3.
Illumina/Solexa sequencing approach. (1) The DNA templates with the attached adapter sequences are connected via a glass surface coated with oligos complementary sequences (2, 3, 4). DNA molecules fold over into a bridge shape and bridge PCR amplification is applied. (5) Bridge amplification and the formation of millions of copies or cluster formation. (6) Cluster sequencing is achieved through the process of cyclic reversible termination method. Finally, the resulting reads (tens of millions) are analyzed and the DNA sequence is recoded.

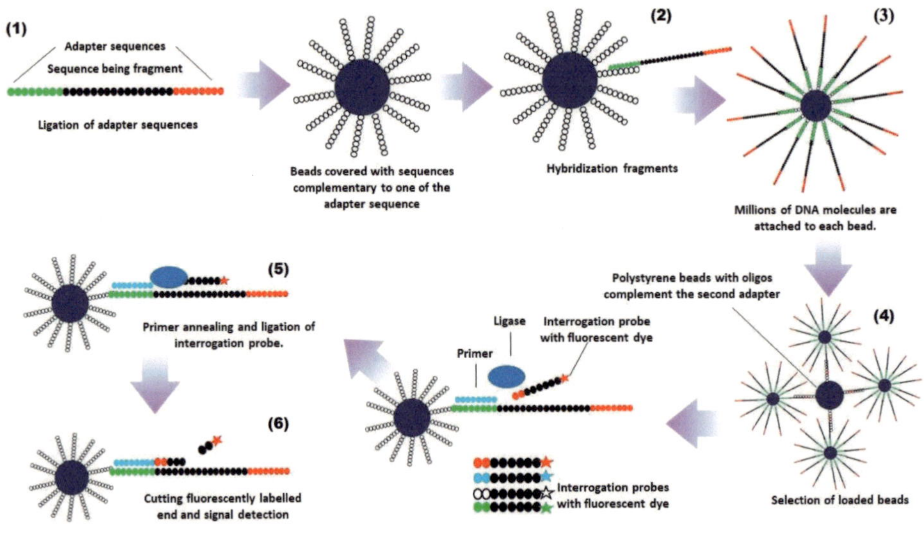

Figure 4.
Applied biosystems (AB) SOLiD sequencing approach. (1) Preparation of DNA library from the sample and ligation of specific adaptor and the beads are then covered with the sequences complementary to one of the adapter sequences. (2) The adapter sequences will then bind to its complementary sequences on the beads. (3) The hybridization process resulted in the attachment of millions of DNA sequences to the bead. (4) Removal of the unloaded beads and selection of the loaded beads. (5) An interrogation probe contains six universal bases and two-base encoded probe. The universal bases are attached to the fluorescent label. (6) When an integrated probe is ligated with primers using DNA ligase, fluorescent light is generated and detected. This process is repeated several times till the targeted DNA is completely sequenced.

two next-generation sequencing technologies, Illumina, and 454 pyrosequencing. AB SOLiD platform relies on sequencing-by-oligo-ligation (SBL) (**Figure 4**), whereas others rely on sequencing-by-synthesis (SBS) [33]. In SOLiD sequencer, the DNA library is prepared from the sample, and specific adaptor is then amplified by emPCR [34]. Instead of utilizing DNA polymerase, short nucleotides marked by DNA ligase known as interrogation probes are used. The interrogation probe contains six universal bases and two-base encoded probe. The universal bases are attached to the fluorescent label. When an integrated probe is ligated with primers using DNA ligase, fluorescent light is generated and detected. After the 5′ end that is linked to the fluorescent label by cleavable linkage is cleaved and removed, thereby the next interrogation probe is connected. This process is repeated several times until the targeted DNA is completely sequenced. In fact, the read length of SOLiD is short about 85 bp leading to inaccurate read assembly as it requires more time for sequencing but it has the highest accuracy among other NGS [35]. Application of SOLiD includes whole genome sequencing, targeted sequencing, transcriptome, and epigenome [35].

2.2.1.4 Ion torrent sequencing

Ion Torrent has been launched in 2010 by Life Technology. Some authors have classified the Ion Torrent platform as a technique between the next-generation and the third-generation sequencing. This could be attributed to the dependence of this approach on optical sensors. However, it relies on chemical sensors that detect the hydrogen-ion concentration change that occurred during the incorporation of a nucleotide in the sequence [21]. Ion Torrent sequencing quality is high and stable due to the utilizing of a chemical sensor instead of fluorescence and camera. In addition, the Ion Torrent approach is characterized by its high speed and low cost compared with pyrosequencing and Illumina [35].

2.2.2 Third-generation sequencing

The major limitations of NGS are that the short-read length and the PCR bias are introduced by clonal amplification and the fluorescent-based signaling detection [21]. Therefore, the third-generation sequencing or single-molecule-sequencing technologies (SMS) overcome these limitations by dispensing PCR before sequencing, and the signal is captured in real time by monitoring the enzymatic reaction [36, 21]. The following sections discuss some TGS platforms.

2.2.2.1 Helicos biosciences (HeliScope)

The first single-molecule-sequencing (SMS) that has been introduced in 2008 is HeliScope. It is a fluorescent-based, single-molecule-sequencing platform. In HeliScope platform, the preparation step depends on preparing a single-strand DNA, and there is no need for PCR amplification in the preparation step. During sequencing, repetitive cycles of DNA polymerase and one labeled nucleotide are flowed, resulting in DNA template extension which depends on the flow of nucleotides. The labeled nucleotides are modified by attaching a poly-A tail in order to stop polymerase extension until the fluorescence that generates from the incorporated nucleotide is recorded by a CCD camera. Then unincorporated nucleotides are washed out and the fluorescent labels on the strand chemically removed, allowing for next base incorporation [37, 38]. HeliScope Genetic Analysis System platform allows the sequencing of RNA, and there is no need for converting them to cDNA. Furthermore, HeliScope Genetic Analysis System platform is in its infancy due to small read length (24–70 bases) and low data output (20 GB) [39].

2.2.2.2 PacBio technology/SMRT sequencer

Pacific Bioscience has launched a single-molecule real-time (SMRT) technology in 2010. It is a real-time, fluorescent-based, and single-molecule-sequencing platform. In SMRT, there is no need for PCR amplification during DNA preparation [36]. In this platform, a nanostructure known as zero-mode waveguide (ZMW) is utilized for real-time observation of DNA synthesis. During the sequencing process, a single-stranded template is used to synthesize the complementary. Unlike other NGS platforms, four different colored fluorescent labels are attached to the terminal phosphate group instead of attaching to a nucleotide, resulting in the release of a fluorescent signal during nucleotide incorporation [40]. Then the camera captures the fluorescent signal in real time (like a movie) [41]. In SMRT, the washing step between nucleotide flows is not required, resulting in increasing the nucleotide incorporation and improving the quality of sequencing [42]. SMRT has several advantages including fast sample preparation (hours instead of days like NGS), no need for PCR amplification during the preparation step, and longer-read length than any other next-generation sequencing platform [42].

2.2.2.3 Oxford Nanopore technology

Nanopore sequencing, developed by Oxford Nanopore Technology, relies on passing the DNA sequence through 1 nm diameter hole (nanopore) where electric current is applied. The electrical current of the pore is altered for each nucleotide, and signal is detected in real time [39]. Like other third-generation sequencing approaches, this technology does not require PCR amplification or chemical labeling of the sample [43]. In May 2015, Oxford Nanopore Technologies has introduced commercially the MinION. The MinION is a pocket-size portable, real-time detection of bases (fluorescent tag-free), has long-read length, and is a low-cost technology [44, 41, 45]. Interestingly, by utilizing this technology, samples can be sequenced in the field directly, instead of collecting samples and sequencing them in the lab, which means nanopore sequencing will make all other sequencing machines redundant [46, 44].

2.3 Metagenomic data analysis

Several bioinformatic tools were developed to analyze the metagenomic data at the molecular level (e.g., 16S rRNA), species level, and strain level. 16S rRNA sequence strategy is among the most common approaches to understand microbial taxonomy and phylogeny. This could be attributed to the stable functions of 16S rRNA gene over time, the existence of 16S rRNA in nearly all microorganisms, and its size which is enough for bioinformatics analysis [47, 48]. A number of bioinformatics tools are available for the analysis of 16S rRNA: QIIME, MOTHUR, DADA2, UPARSE, and minimum entropy decomposition (MED) [49]. The QIIME software is designed to analyze data generated on the Illumina or other NGS platforms via graphics and statistics. This involves the demultiplexing and quality filtering, OTU picking, taxonomic assignment, and phylogenetic reconstruction, and diversity analyses and visualizations [50, 51]. QIIME depends on the use of the PyCogent toolkit to identify misinterpretations and database deposition using raw sequencing results [51]. Operational taxonomic units (OTUs) can be generated from NGS data by UPARSE [52]. The UPARSE software acts by filtering and trimming reads into equals lengths, removing singleton reads and clustering the remaining reads [52].

Community sequence data can be analyzed by a flexible and comprehensive software package called MOTHUR. The MOTHUR package includes the following algorithms: DOTUR, SONS, TreeClimber, LIBSHUFF, Đ-LIBSHUFF, and UniFrac [50]. DADA2 is a suitable approach for correcting amplicon errors with no option to generate OTUs [53]. DADA2 uses a new quality-aware model of Illumina amplicon errors to improve the DADA algorithm [53]. MED is applied to solve the limitations of fine-scale resolution descriptions of microbial communities [54]. MED acts through partitioning the data set of amplicon sequences into homogenous OTUs for alpha- and beta-diversity analyses [54].

For species-level metagenomic data analysis, there are at least six metagenomic analysis software including MetaPhlAn2 [55], Kraken [56], CLARK [57], FOCUS [58], SUPERFOCUS [59], and MG-RAST [60]. All of these software programs can be used to profile organisms in metagenomic samples and to score their abundance. MetaPhlAn2 applies Bowtie2 and UCLUST [52, 61] as its main algorithms, whereas k-mers (DNA words of length k) is the core algorithm for Kraken and CLARK. On the other hand, FOCUS uses the NNLS (nonnegative least squares) to identify the microbial profile [49].

3. Conclusion

At the beginning, the metagenomic workflow was complicated that it requires many steps, sophisticated equipment, and qualified technicians to perform. Likewise, it was very expensive that not all scientists or laboratories were able to afford its cost. However, nowadays, due to the presence of many different competing companies and laboratories that led to the development of more efficient sequencing approaches, the metagenomic workflow became easier. It is easy now to study and identify organisms directly from their habitats without prior preparations. In terms of cost, NGS is also much cheaper, and with the appearance of third-generation sequencing approaches, it is not required to conduct sample sequencing. Surprisingly, sequencing can be carried out in the field by utilizing a pocket-size portable sequencer. The advancements in the field of metagenomics are amazing, and it became easier, cheaper, and faster.

Acknowledgements

This work was funded by the Deanship of Scientific Research at Princess Nourah Bint Abdulrahman University, through the Research Groups Program Grant no. RGP-1438-0004.

Conflict of interest

The authors declare that there are no conflicts of interest.

Author details

Ahmed Shuikan[1], Sulaiman Ali Alharbi[1], Dalal Hussien M. Alkhalifah[2] and Wael N. Hozzein[3,4*]

1 Botany and Microbiology Department, College of Science, King Saud University, Riyadh, Saudi Arabia

2 Biology Department, Faculty of Science, Princess Nourah Bint Abdulrahman University, Riyadh, Saudi Arabia

3 Bioproducts Research Chair, Zoology Department, College of Science, King Saud University, Riyadh, Saudi Arabia

4 Botany and Microbiology Department, Faculty of Science, Beni-Suef University, Beni-Suef, Egypt

*Address all correspondence to: hozzein29@yahoo.com; whozzein@ksu.edu.sa

References

[1] Xu J. Microbial ecology in the age of genomics and metagenomics: Concepts, tools, and recent advances. Molecular Ecology. 2006;**15**:1713-1731

[2] Schlaberg R, Chiu CY, Miller S, Procop GW, Weinstock G. Microbiology Resource Committee of the College of American Pathologists. Professional Practice Committee and Committee on Laboratory Practices of the American Society for Microbiology, et al. Validation of metagenomic next generation sequencing tests for universal pathogen detection. Archives of Pathology & Laboratory Medicine. 2017;**141**:776-786

[3] Escobar-Zepeda A, Vera-Ponce de León A, Sanchez-Flores A. The road to metagenomics: From microbiology to DNA sequencing technologies and bioinformatics. Frontiers in Genetics. 2015;**6**:348

[4] Almeida OGG, DeMartinis OCP. Bioinformatics tools to assess metagenomic data for applied microbiology. Applied Microbiology and Biotechnology. 2019;**103**(1):69-82

[5] Greninger AL. The challenge of diagnostic metagenomics. Expert Review of Molecular Diagnostics. 2018;**18**:605-615

[6] Gu W, Miller S, Charles Y. Chiu. Clinical metagenomic next-generation sequencing for pathogen detection. Annual Review of Pathology. 2019;**14**:319-338

[7] Greninger AL, Zerr DM, Qin X, Adler AL, Sampoleo R, Kuypers JM, et al. Rapid metagenomic next generation sequencing during an investigation of hospital-acquired human parainfluenza virus 3 infections. Journal of Clinical Microbiology. 2017;**55**:177-182

[8] Simner PJ, Miller HB, Breitwieser FP, Monsalve GP, Pardo CA, Salzberg SL, et al. Development and optimization of metagenomic next-generation sequencing methods for cerebrospinal fluid diagnostics. Journal of Clinical Microbiology. 2018;**56**:e00472-e00418

[9] Vincent AT, Derome N, Boyle B, Culley AI, Charette SJ. Next-generation sequencing (NGS) in the microbiological world: How to make the most of your money. Journal of Microbiological Methods. 2017;**138**:60-71

[10] Quince C, Walker AW, Simpson JT, Loman NJ, Segata N. Shotgun metagenomics, from sampling to analysis. Nature Biotechnology. 2017;**35**(9):833

[11] Young JM, Rawlence NJ, Weyrich LS, Cooper A. Limitations and recommendations for successful DNA extraction from forensic soil samples: A review. Science & Justice. 2014;**54**(3):238-244

[12] Finley SJ, Lorenco N, Mulle J, Robertson BK, Javan GT. Assessment of microbial DNA extraction methods of cadaver soil samples for criminal investigations. Australian Journal of Forensic Sciences. 2016;**48**(3):265-272

[13] Lim NY, Roco CA, Frostegård Å. Transparent DNA/RNA co-extraction workflow protocol suitable for inhibitor-rich environmental samples that focuses on complete DNA removal for transcriptomic analyses. Frontiers in Microbiology. 2016;**7**:1588

[14] Gupta P, Manjula A, Rajendhran J, Gunasekaran P, Vakhlu J. Comparison of metagenomic DNA extraction methods for soil sediments of high elevation Puga hot spring in Ladakh, India to explore bacterial diversity. Geomicrobiology Journal. 2017;**34**(4):289-299

[15] Mazziotti M, Henry S, Laval-Gilly P, Bonnefoy A, Falla J. Comparison of two

bacterial DNA extraction methods from non-polluted and polluted soils. Folia Microbiologica. 2018;**63**(1):85-92

[16] Van Djick EL, Auger H, Jaszczyszyn Y, Thermes C. Ten years of next-generation sequencing technology. Trends in Genetics. 2014;**30**:418-426

[17] Head SR, Komori HK, LaMere SA, Whisenant T, Van Nieuwerburgh F, Salomon DR, et al. Library construction for next generation sequencing: Overviews and challenges. Biotech. 2014;**56**:61-68

[18] Simon C, Daniel R. Construction of small-insert and large-insert metagenomic libraries. In: Metagenomics. New York, NY: Humana Press; 2017. pp. 1-12

[19] Sanger F, Coulson AR. A rapid method for determining sequences in DNA by primed synthesis with DNA polymerase. Journal of Molecular Biology. 1975;**94**:441-448

[20] Maxam AM, Gilbert W. A new method for sequencing DNA. Proceedings of the National Academy of Sciences. 1977;**74**:560-564

[21] Schadt EE, Truner S, Kasarskis A. A window into third-generation sequencing. Human Molecular Genetics. 2010;**19**:R227-R240

[22] Sanger F, Nicklen S, Coulson AR. DNA sequencing with chain-terminating inhibitors. Proceedings of the National Academy of Sciences. 1977;**74**:5463-5467

[23] Hert DG, Fredlake CP, Barron AE. Advantages and limitations of next-generation sequencing technologies: A comparison of electrophoresis and non-electrophoresis methods. Electrophoresis. 2008;**29**:4618-4626

[24] Venter JC, Adams MD, Myers EW, Li PW, Mural RJ, Sutton GG, et al. The sequence of the human genome. Science. 2001;**291**:1304-1351

[25] Metzker ML. Sequencing technologies-the next generation. Nature Reviews. Genetics. 2010;**11**:31-46

[26] Varshney RK, Nayak SN, May GD, Jackson SA. Next-generation sequencing technologies and their implications for crop genetics and breeding. Trends in Biotechnology. 2009;**27**:522-530

[27] Sogin ML, Morrison HG, Huber JA, Mark Welch D, Huse SM, Neal PR, et al. Microbial diversity in the deep sea and the underexplored 'rare biosphere'. Proceedings of the National Academy of Sciences. 2006;**103**:12115-12120

[28] Mardis ER. Next-generation DNA sequencing methods. Annual Review of Genomics and Human Genetics. 2008;**9**:387-402

[29] Margulies M, Egholm M, Altman W, et al. Genome sequencing in microfabricated high-density picolitre reactors. Nature. 2005;**437**:376-380

[30] Huse SM, Huber JA, Morrison HG, Sogin ML, Welch DM. Accuracy and quality of massively parallel DNA pyrosequencing. Genome Biology. 2007;**8**(7):R143

[31] Fullwood MJ, Wei CL, Liu ET, Ruan Y. Next-generation DNA sequencing of paired-end tags (PET) for transcriptome and genome analyses. Genome Research. 2009;**19**(4):521-532

[32] Glenn TC. Field guide to next generation DNA sequencers. Molecular Ecology Resources. 2011;**11**(5):759-769. Available from: https://www.molecularecologist.com/next-gen-fieldguide-2016/

[33] McKernan K, Blanchard A, Kotler L, Costa G. Reagents, methods

and libraries for bead-based sequencing. US20080003571; 2008

[34] Shao K, Ding W, Wang F, Li H, Ma D, Wang H. Emulsion PCR: A high efficient way of PCR amplification of random DNA libraries in aptamer selection. PLoS One. 2011;**6**(9):e24910. DOI: 10.1371/journal.pone.0024910

[35] Liu L, Li Y-H, Li S, Hu N, He Y, Pong R, et al. Comparison of next-generation sequencing systems. Journal of Biomedicine & Biotechnology. 2012;**2012**:251364. DOI: 10.1155/2012/251364

[36] Korlach J, Bjornson KB, Chaudhuri BP, Cicero RL, Flusberg BA, Gray JJ, et al. Real-time DNA sequencing from single polymerase molecules. Methods in Enzymology. 2010;**472**:431-455

[37] Harris T, Buzby P, Babcock H, et al. Single molecule DNA sequencing of a viral genome. Science. 2008;**320**:106-109

[38] Zhang J, Chiodini R, Badr A, Zhang G. The impact of next-generation sequencing on genomics. Journal of Genetics and Genomics. 2011;**38**:95-109

[39] Hart C, Lipson D, Ozsolak F, Raz T, Steinmann K, Thompson J, et al. Single molecule sequencing: Sequence method to enable accurate quantitation. Methods in Enzymology. 2010;**472**:407-430

[40] Flusberg BA, Webster DR, Lee JH, Travers KJ, Olivares EC, Clark TA, et al. Direct detection of DNA methylation during single-molecule, real-time sequencing. Nature Methods. 2010;**7**:461-465

[41] Timp W, Mirsaidov UM, Wang D, Comer J, Aksimentiev A, Timp G. Nanopore sequencing: Electrical measurements of the code of life. IEEE Transactions on Nanotechnology. 2010;**9**(3):281-294

[42] Zhou X, Ren L, Li Y, Zhang M, Yu Y, Yu J. The next-generation sequencing technology: A technology review and future perspective. Science China. Life Sciences. 2010;**53**:44-57

[43] Shi Y, Tyson GW, Eppley JM, DeLong EF. Integrated metatranscriptomic and metagenomic analyses of stratified microbial assemblages in the open ocean. The ISME Journal. 2011;**5**:999-1013

[44] Hayden EC. Nanopore genome sequencer makes its debut. Nature. 2012;**10051**. DOI: 10.1038/nature.2012.10051

[45] Laver T, Harrison J, O'Neill PA, Moore K, Farbos A, Paszkiewicz K, et al. Assessing the performance of the Oxford Nanopore technologies MinION. Biomolecular Detection and Quantification. 2015;**3**:1-8

[46] Jain M, Fiddes IT, Miga KH, Olsen HE, Paten B, Akesen M. Improved data analysis for the MinION nanopore sequencer. Nature Methods. 2015;**12**:351-356

[47] Patel JB. 16S rRNA gene sequencing for bacterial pathogen identification in the clinical laboratory. Molecular Diagnosis. 2001;**6**:313-321

[48] Janda JM, Abbott SL. 16S rRNA gene sequencing for bacterial identification in the diagnostic laboratory: Pluses, perils, and pitfalls. Journal of Clinical Microbiology. 2007;**45**:2761-2764

[49] Niu SY, Yang J, McDermaid A, Zhao J, Kang Y, Ma Q. Bioinformatics tools for quantitative and functional metagenome and metatranscriptome data analysis in microbes. Briefings in Bioinformatics. 2018;**19**(6):1415-1429. DOI: 10.1093/bib/bbx051

[50] Schloss PD, Westcott SL, Ryabin T, et al. Introducing MOTHUR: Open-source, platform-independent,

community supported software for describing and comparing microbial communities. Applied and Environmental Microbiology. 2009;**75**:7537-7541

[51] Caporaso JG, Kuczynski J, Stombaugh J, et al. QIIME allows analysis of high-throughput community sequencing data. Nature Methods. 2010;**7**:335-336

[52] Edgar RC. UPARSE: Highly accurate OTU sequences from microbial amplicon reads. Nature Methods. 2013;**10**:996-998

[53] Callahan BJ, McMurdie PJ, Rosen MJ, Han AW, Johnson AJA, Holmes SP. DADA2: High resolution sample inference from Illumina amplicon data. Nature Methods. 2016;**13**:581-583

[54] Eren AM, Morrison HG, Lescault PJ, Reveillaud J, Vineis JH, Sogin ML. Minimum entropy decomposition: Unsupervised oligo typing for sensitive partitioning of high-throughput marker gene sequences. The ISME Journal. 2015;**9**:968-979

[55] Truong DT, Franzosa EA, Tickle TL, Scholz M, Weingart G, Pasolli E, et al. MetaPhlAn2 for enhanced metagenomic taxonomic profiling. Nature Methods. 2015;**12**:902-903

[56] Wood DE, Salzberg SL. Kraken: Ultrafast metagenomic sequence classification using exact alignments. Genome Biology. 2014;**15**:R46

[57] Ounit R, Wanamaker S, Close TJ, Lonardi S. CLARK: Fast and accurate classification of metagenomic and genomic sequences using discriminative k-mers. BMC Genomics. 2015;**16**:236

[58] Silva GG, Cuevas DA, Dutilh BE, Edwards RA. FOCUS: An alignment free model to identify organisms in metagenomes using non-negative least squares. Peer J. 2014;**2**:e425. DOI: 10.7717/peerj.425

[59] Silva GG, Green KT, Dutilh BE, Edwards RA. SUPER-FOCUS: A tool for agile functional analysis of shotgun metagenomic data. Bioinformatics. 2016;**32**:354-361

[60] Meyer F, Paarmann D, D'Souza M, Olson R, Glass EM, Kubal M, et al. The metagenomics RAST server-a public resource for the automatic phylogenetic and functional analysis of metagenomes. BMC Bioinformatics. 2008;**9**:386

[61] Langmead B, Salzberg SL. Fast gapped-read alignment with bowtie 2. Nature Methods. 2012;**9**:357-359

Chapter 4

The Use of Bioinformatic Tools in Symbiosis and Co-Evolution Studies

Raúl Enrique Valle-Gough,
Blancka Yesenia Samaniego-Gámez,
Javier Eduardo Apodaca-Hernández
and Maria Leticia Arena-Ortiz

Abstract

Through millions of years, the multicellular organisms have coexisted and coevolved with the surrounding microorganisms, in an almost symbiotic relationship forming a complex entity known as holobiont. The composition and functions of these microbial communities were limited during many years to only a mere fraction, due to the use of culture-based techniques. The advent of molecular-based techniques allowed the identification of uncultured organisms in a culture-free manner. In recent years, the development of next generation sequencing techniques have allowed the high-throughput study of microbial communities allowing the identification and classification of otherwise uncultured microorganisms in a given environment, tissue or host through metagenomics. The next generation sequencing techniques have been used in the functional study of microbial assemblages and were able to identify the role of the microorganisms in biogeochemical cycles, pathogenic processes, metabolism and development, through metatranscriptomics. Taken together, the next generation sequencing based-studies have shown the existence of a complex metabolic network in different hosts and environments, with the microbial communities. This chapter will focus in different available bioinformatic tools that are suitable to study symbiosis and coevolution processes in a given sample.

Keywords: next generation sequencing, bioinformatics, metagenomics, metatranscriptomics, network analysis

1. Introduction

Higher multicellular organisms have coexisted and co-evolved with resident microorganisms in a relatively harmonious relationship over millions of years, forming a complex organism called holobiont. These processes of co-evolution have been documented by several studies carried out in the organization and composition of host microorganisms (microbiome) in different species [1–4]. The microbiome is currently considered a functional organ which is fundamental for the host organism, given that studies have shown that this organ is highly dynamic and adaptable, likewise plays an important role in physiological adaptation processes, metabolism, and development [1, 2, 4–6]. The study of the role of the microbiome for years was limited to those organisms that could be susceptible to culture, the use of techniques based on molecular information (AFLP and RFLP) and denaturing gradient gel electrophoresis (DGGE) could reveal the presence of some species not cultured, but the gel resolution was being the main limitation since a single band could contain more than one sequence [7]. The development of genomic tools such as the new generation sequencing platforms (NGS) has allowed a better resolution of the diversity present in a given sample through what is known as metagenomics [2, 7–10]. These same sequencing platforms have allowed not only observing the diversity of a host or environment, but also the functional role of microorganisms as well as their possible interactions with physiological processes or biogeochemical cycles through metatranscriptomics [1, 2, 4, 9]. The understanding of the composition of microorganisms, functions, interactions, and other biological processes through NGS has been done through the development of different bioinformatics tools, which make use of large amounts of information and are able to compare them through different data bases [3, 9, 11]. The objective of this chapter is to provide information on the different existing bioinformatics tools that have been used in studies of co-evolution and symbiosis in different models.

2. Bioinformatic tools in metagenomics and metatranscriptomics in different samples

The first step of a NGS-based study involves the extraction of nucleic acids in sufficient quantity and quality to carry out the sequencing process in order to have an unbiased knowledge of the microbial diversity present in a sample [6, 12, 13]. The processing of DNA samples (environmental and host) can be performed by cell recovery by centrifugation gradients in differential media and the subsequent recovery of DNA by silica columns [6]. Another methodology used is the *in situ* lysis of the sample by the addition of enzymes (Proteinase K and lysozyme) with the subsequent separation of cell debris by centrifugation and recovery of DNA by solvent precipitation or by silica adsorption [14]. The main advantage of *in situ* lysis is that higher amounts of DNA are obtained when compared to cell recovery techniques; however, there is a risk of the presence of contaminants that may interfere with sequencing reactions [6].

In the case of RNA, the main methodologies perform *in situ* lysis of the sample under RNase-free conditions using different guanidine solvents and salts to avoid the presence of ribonucleases [15]. The samples should be placed at −80°C either in dry ice or liquid nitrogen to avoid their degradation.

Quality control of nucleic acids can be carried out by visualization on agarose gels, by spectrophotometric means (Nanodrop) and in microfluidic chambers (Bioanalyzers). This last system has been widely used since it allows the visualization and simultaneous quantification of nucleic acids [8, 16–18]. In the case of RNA, these systems have developed a scale known as RNA Integrity Number (RIN) which, based on the proportion between the major and minor subunits of the rRNA assigns a minimum value that must be greater than 8.0.

2.1 NGS shared tools for metagenomics and metatranscriptomics

Once a sample with sufficient quality and quantity was sent to sequencing, a series of files with the ".fastq" extension are obtained, which contains the information of the sequence and the quality for each base. This format is used by different programs (FASTQC and PRINSEQ) to perform the quality control of the sequencing, showing basic statistics such as the total number of bases, read size, GC content, quality for each base in PHRED33 or PHRED64 scale, as well as the presence of overrepresented sequences [8, 19–23]. The files analyzed are introduced to different programs (Trimmomatic, TrimGalore, and CutAdapt) that trims the reads of the ".fastq" file, based on the quality for each nucleotide, eliminating sequences with a PHRED value below 20 and a minimum fragment size defined by the user [19, 20, 22, 24].

These programs are able to eliminate segments of initiators and sequencing adapters, which must be provided in a separate file. The output files of these programs are archives in ".fastq" format, where the sequences that are common for all samples are placed in one file, and the unique sequences for each individual sample are placed in several files [19, 20].

2.2 Metagenomic tools used in symbiosis and co-evolution studies

In recent years, the use of genomic approaches has revealed an unprecedented diversity and bacterial ubiquity in different types of samples (**Figure 1**), through the analysis of 16S ribosomal sequences [1, 2, 5, 6, 18, 19, 22, 25–27]. These techniques have allowed the molecular analysis of populations and how different biological processes have been established, controlled, and evolved [5, 28, 29].

The metagenomic composition analyzes have been carried out through the use of different programs (QIIME, QIIME2, and MOTHUR), that align the reads against a database of ribosomal genes (GreenGenes, SILVA, and RDP) and assign them operational taxonomic units (OTUs), using a distance of 3% and a confidence interval of 80% [29–33]. Once the OTUs have been assigned, the aforementioned programs allow the determination of diversity indices, richness, and main component analysis and perform the rarefaction of the samples [1, 2, 5, 19, 25, 29–32, 34, 35].

Other taxonomic classifiers are based on alignment of short sequences previously edited, by single or paired ends (Kraken, Kraken2, OneCodex) comparing them with the databases available in each program. In the case of Kraken, it makes

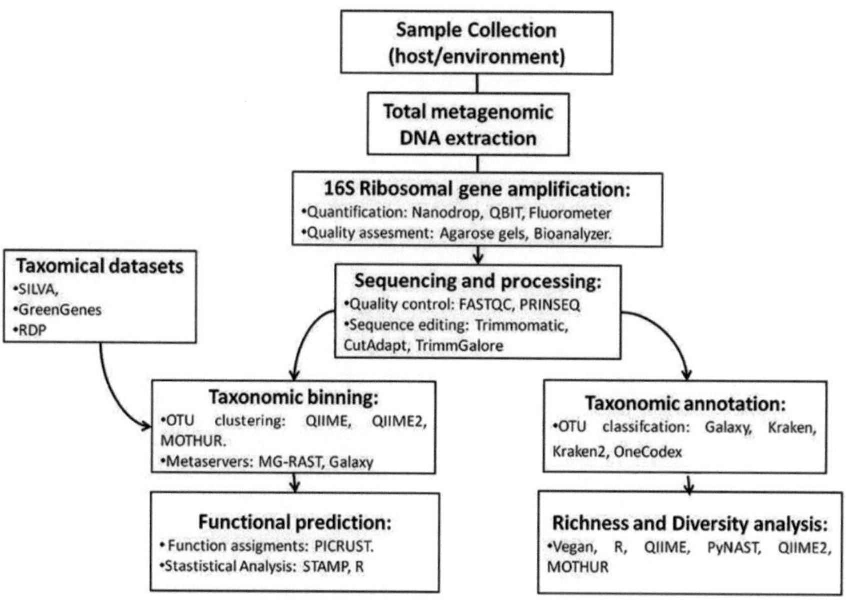

Figure 1.
Genomic and metagenomic techniques for the analysis in different samples.

use of the Ref-Seq database where the reads are divided into fragments known as k-mers and are compared with sequenced genomes [34, 36]. The resulting files of these programs are provided in tabular format (tsv), which facilitates their export and processing in other types of programs such as Vegan or R, where studies of richness, diversity, and rarefaction can be carried out [12, 34, 35, 37, 38].

The use of different taxonomic binning programs has been able to determine the presence of ubiquitous microbial phyla present in samples from arctic, temperate, and tropical environments such as: *Proteobacteria, Actinobacteria*, and *Cyanobacteria*, which are considered cosmopolite phyla. The main difference between each site is the proportion of each taxa, which reflects the conditions of each environment [6, 8–10, 39–41]. A similar behavior has been observed when studying the microbiome in different animal models where the phyla: *Proteobacteria, Acitnobacteria, Firmicutes*, and *Bacteroidetes* have been reported among those of greater relative abundance [4, 9, 39, 42–46]. This shows that microbial communities are highly dynamic where the physical-chemical factors of the site, health status, and nutrition shape the metagenome and can determine how reactive a microbial community is to environmental changes.

The use of genomic tools has made possible to identify the core microbiome of different organisms, given that, despite living in different habitats, they share similar bacterial communities, which implies the existence of biological filters that shape the bacterium-host interactions, resulting in a stable relationship with the holobiont [2, 28, 45–47]. In the case of *Apis mellifera*, a global core microbiome formed by *Proteobacteria, Firmicutes, Bacteroidetes,* and *Actinobacteria* has been identified, together with a high amount of lactic acid bacteria which have a beneficial activity in the health of the host organism due to their involvement in the immunomodulation of the intestinal microflora [25, 39, 48]. The presence of symbiont microorganisms within the intestinal tract in different animal species (*A. mellifera*, *Litopenaeus vannamei, Mus musculus, Homo sapiens*) have been reported as necessary for survival, since their cooperative behavior increases the vigor of a community [28, 39, 47, 49, 50]. Recent studies in fecal samples of farm animals have revealed the presence of intervening sequences (IVS), which are host-specific and provide a basis for the differentiation of the microorganisms derived from different hosts [51].

The role of microbial communities within a host is important. Given the existing delicate balance of these associations, any type of alterations in the microbiome composition could cause disease in the host organism [6, 12, 39, 45]. Previous studies have revealed that in diseased individuals of different species, the microbial diversity is significantly reduced. This could be due to the fact that alterations in the microbiome composition skew the association between the host and the microbiome producing dysbiosis and increasing the number of opportunistic pathogens [6, 12, 13, 16, 27, 39, 45, 52]. In marine environments, the continual presence of pathogens has been observed in environmental samples [16, 44] and in several marine organisms (*L. vannamei* and *M. nipponense*) [17, 18, 39, 45, 46]. The continual presence of pathogens in low proportion has been reported during the life cycle of these species, suggesting an active *in situ* infection in which the host has co-evolved with the parasitic organisms and developed mechanisms that cope with the pathogenic mechanisms of the parasites [13, 16, 17, 27, 45, 46, 52]. It has been observed that the developmental stage in *L. vannamei* influences the pathogenic response to *Vibrio*, where the proportion of protective commensal bacteria, *Bacteroides* and *Propionibacterium*, tend to decrease as the host aged in contrast the presence of *Vibrio* increases in diseased individuals [18]. Other mechanisms of coevolution have shown that processes of parasitism and predation can influence the global exchange of resources in an ecosystem. Studies conducted in *Escherichia coli* and the bacterial predator *Myxococcus xanthus* have shown that the genome evolution of the predator and prey exhibited accelerated genome evolution when compared to controls, where the predator (*M. xanthus*) showed adaptations to cell mucoidy and the prey (*E. coli*) showed adaptations to outer membrane-proteases [7].

The functional analysis of the microbial communities has been carried out using the PICRUST program. This program estimates the families of genes present in a metagenome, by the phylogenetic comparison with sequences of gene families previously reported in databases. These predictions are pre-calculated for genes that code for proteins present in orthologous gene families (COG) or in the Kyoto Encyclopedia of Genes and Genomes (KEEG) [53]. The differential expression of these predicted functions could be assessed with the STAMP software which allows several statistical analysis, size effect, and sample corrections [54]. The use of the afore mentioned protocols have allowed the observation of various attributes in environmental samples related with carbon fixation, amino acid metabolism, and signal transduction in lakes, swamps, and other water bodies [9, 10, 16, 22, 44, 55]. These reports also showed the presence of several bacterial taxa (*Actinobacteria, Verrucomicrobia*, and *Proteobacteria*) who were able to synthesize several extracellular enzymes that digests the organic matter [9, 16, 24] or mineralize other nutrients [22, 44].

The influence of the microbiome on the host function have been proposed as a co-evolutionary process where the functionality and the composition of the microbiome can be influenced by the feeding habits of the host [4, 21], and the host can take advantage of the specialized microorganisms who are able to synthesize metabolites that are not present originally in the environment [6, 39]. The consumption of seaweeds by Japanese allows the introduction of algae associated bacteria, which transfer the genes involved in the degradation of the algal sulphated polysaccharides to competent gut resident bacteria with a process known as horizontal gene transfer [28]. Certain marine invertebrates (*Elysia chlorotica*) that feed on algae are able to maintain the algal plastids as photosynthetically symbionts which allow the use of photosynthates as food source [26]. These examples of coevolutionary processes show how the functionality of the microbiome could be influenced by the dietary habits of the host since; these metabolic add-ons allow the host to thrive in otherwise adverse environmental conditions (oligotrophic habitats).

2.3 Metatranscriptomic tools used in symbiosis and co-evolution studies

The metatranscriptomic allows the establishment of parallel relations between the host and the microbiome, but studies require a series of previous steps in order to obtain unbiased information such as the removal of rRNA and the microbial mRNA enrichment (**Figure 2**) [17, 19–21, 56].

The assembly of genomes and transcriptomes uses short sequences that are separated into fragments known as *k-mers*, which are aligned and compared graphically (De Brujin graphs) in order to perform *de novo* reconstruction of the genome or transcriptome. Several programs such as Velvet, SOAP, Trinity, and FLASH are capable of performing it by using a reference genome or transcriptome, if available [8, 57–63]. In the case of the Trinity platform, it is capable not only of assembling but also of mapping within the assembly (Bowtie1 and Bowtie2), basic statistical analysis of the assembly, quantifying transcripts (RSEM, Salmon, eXpress, and Kallisto), and performing differential expression of transcripts (edgeR, DESeq2, ROTS, and lima/voom) [8].

The metatranscriptomic studies have allowed to reveal the functions of the microorganisms within a host or in different environments and to identify, in both host and microbiome, transcripts related mainly to metabolic processes associated with the nutrient uptake. These observations suggest that the symbiotic chemoautotrophic bacteria provide organic compounds to the host organism that uses it for

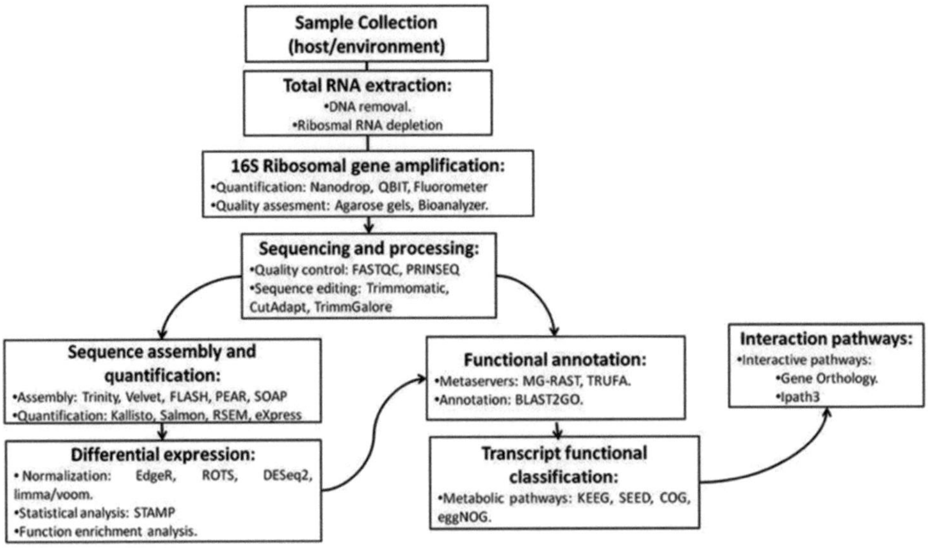

Figure 2.
Transcriptomic and metatranscriptomic techniques for the analysis in different samples.

its nutrition [11, 20, 45, 64]. In fact, recent studies have reported that more than a third of genes are shared among living organisms, especially to those related to the central metabolic pathways (Glucolysis, TCA, Oxidative phosphorylation, Purine and Pyrimidine metabolism) which could increase the efficiency for the digestion of several biomolecules [11, 17, 26, 45].

The use of bioinformatic tools in metatranscriptomics studies has allowed the visualization of the host-microbiome interactions, especially those related with the primary metabolism [13, 38, 52]. The visualization of the shared enzymatic modules is accomplished through the use of identifiers derived from KEGG orthology (KO) and Enzyme Codes (EC) on the iPath3 platform [65]. In this platform, it is possible to overlap metabolic functions (host-symbiont) using the EC and KO identifiers in different metabolic maps (general metabolic pathways, bacterial metabolism, and secondary metabolism), showing graphically the enzymatic modules of each individual and highlighting the enzymatic modules with a shared function.

Metatranscriptomic studies have been able to show that microorganisms are capable of generating complex trophic networks communicating with each other through chemical signals in a process known as quorum sensing [12, 26, 56]; however, this process is not restricted only to microorganisms; recent studies have suggested an interdominion quorum sensing [4, 17, 21].

2.4 Metaservers used in the metagenomic and metatranscriptomics studies

The bioinformatic tools mentioned in this chapter are open-source programs, requiring the user to have a UNIX or OSx operating system installed; a RAM memory greater than 16 GB, a hard disk greater than 500 GB of storage and knowledge about command lines in UNIX [16, 19, 20, 22, 24, 30–32, 35, 46, 48, 51]. These requirements can be complicated for those who want to initiate a bionformatic analysis; however, there are other options, such as the metaservers, which can allow the data processing in a graphical environment.

The metaservers are web service providers that assemble a series of programs and applications that otherwise are dispersed. Among the most used metaservers are Galaxy, TRUFA, and MG-RAST [22, 24, 34].

Galaxy: it is a collaborative initiative that provides a free set of tools and bioinformatics programs ranging from quality control of sequences (FASTQC), sequence editors, data grouping tools, tools for assembly (Trinity), sequence mapping (Bowtie), transcript quantification (Salmon and Kallisto), and metagenomic analysis programs (Mothur, Vegan, Kraken, and Krona) [34]. Being an open initiative, Galaxy presents a series of servers that offer different programs such as the functional prediction of a metagenome by PICRUST (Langille Lab and Huttentowe Lab) and servers dedicated to the functional annotation of transcriptomes (ANASTASIA).

TRUFA: Transcriptome User-Friendly Analysis [22], a program developed by the Institute of Physics of Cantabria, is a free server and contains several programs exclusively for transcriptomic (metatranscriptomic) analysis ranging from quality control (FASTQC and PRINSEQ), edited of sequences (CutAdapt), assembly of sequences (Trinity), quantification of transcripts (RSEM and eXpress) and functional annotation (BLAST2GO and HMMER). The files can be edited beforehand, and certain modules of the platform can be accessed, such as the functional annotation in case of already assembled sequences.

MG-RAST: Metagenomic Rapid Annotation-based on Subsystems Technology [24] is an open platform capable of analyzing sequences from different NGS platforms (Illumina, PacBio, and Nanopore). Unlike the aforementioned servers, MG-RAST has a pipeline that includes the quality control of the sequences, removal of adapters, detection of isoforms of transcripts, taxonomic comparison, and functional assignment. This server has several databases where the results can be analyzed regarding function (SEED, KEEG, COG, and NOG) and taxonomy (ITS, SILVA, RDP, and GreenGenes). It also has tools to export the data in tabular format, fasta, or in the form of BIOM type matrix.

BLAST2GO: it is a sequence annotator that is able to perform searches in the NCBI, which in its basic version has the BLAST algorithms to add taxonomic filters in order to accelerate the annotation. It also allows searches of Interprotein domains (InterProScan), allows the classification of proteins based on the Gene Orthology (GO) database, interaction maps between each GO term, function enrichment analysis (Fisher Exact Test) and the analysis of the metabolic modules present in the KEGG. The PRO version of this program allows making several annotations at the same time, using CLOUD-BLAST services and performing other types of analysis such as the differential expression of transcripts [37].

3. Conclusions

The study of how microbial communities contribute to environmental functions and the physiology of the host was limited to cultivable microorganisms. As, the free culture techniques based on molecular markers developed (AFLP, RFLP, and DGGE), this knowledge expanded. However, the knowledge obtained from these was quite limited. But now with the techniques of massive sequencing, it has been possible to obtain a better understanding of the role of microorganisms in different types of environments and hosts, both from the taxonomic and functional point of view.

The use of bioinformatic programs has allowed not only the reconstruction of the molecular phylogeny but also has allowed them to be studied from the functional point of view, showing the great potential for the future biotechnological exploitation of this microbial metabolic diversity such as enzymes with different catalytic activities from uncultivable organisms. Several combined methodologies that uses NGS techniques along with culture-based methods have been used to obtain new bacterial strains using specific culture media or through the functional screening using specific primers in order to isolate the genes of interest.

Current research is now been carried out in order to obtain more precise metagenomic and metatranscriptomic assemblages, with new software specially designed (MetaVelvet, TriMetAss, and MetaAmos) to obtain complete genomes and transcriptomes of the bacterial communities. These protocols along with the integration of other "omic" techniques and systems biology could allow a better understanding of the complex metabolic and trophic networks that operates in an organism or environment.

Acknowledgements

The authors would like to acknowledge the General Direction of Academic Personal Affairs (DGAPA, UNAM) for the postdoctoral grant for REVG, the National Council of Science and Technology (CONACyT) for funding this project, AquaPacific (Mazatlan, Sinaloa, Mexico) especially to: Bruno Gomez-Gil, Sonia Soto and Daniel Palacios for providing the biological material collection.

Conflict of interest

The author declares that this chapter was conducted in the absence of any financial or commercial relationships that could be construed as a potential conflict of interest.

Author details

Raúl Enrique Valle-Gough[1], Blancka Yesenia Samaniego-Gámez[1], Javier Eduardo Apodaca-Hernández[2] and Maria Leticia Arena-Ortiz[3*]

1 Instituto Tecnológico Superior de Escarcega, Escarcega, Campeche, Mexico

2 Laboratory of Genomic Computing, Genomic Libraries (Bibliotecas Genómicas), Merida, Yucatan, Mexico

3 Universidad Nacional Autonoma de Mexico, Yucatan, Mexico

*Address all correspondence to: leticia.arena@ciencias.unam.mx

References

[1] Aprill A. Marine animal microbiomes: Toward understanding host-microbiome interaction in a changing ocean. Frontiers in Marine Science. 2017;**4**:222. DOI: 10.3389/frmars.2017.00222

[2] Aires T, Yann M, Serrao E, Arnaud-Haond S. Hologenome theory supported by cooccurrence networks of species-specific bacterial communities in siphonous algae (*Caulerpa*). FEMS Microbiology Ecology. 2015;**91**(7):1-14

[3] Degli-Esposti M, Martinez-Romero E. The functional microbiome of arthropods. PLoS One. 2017;**12**(5):e0176573. DOI: 10.1371/journal.pone.0176573

[4] Killian M, Chapple ILC, Hanning M, Marsh PD, Meuric V, Pedersen AML, et al. The oral microbiome- an update for oral healthcare professionals. British Dental Journal. 2016;**221**(10):657-666

[5] Andersen R, Chapmann SJ, Artz RRE. Microbial communities in natural and disturbed peatlands: A review. Soil Biology and Biochemistry. 2015;(57):979-984

[6] Shi W, Syrenne R, Sun JZ, Yuab JS. Molecular approaches to study the insect gut symbiotic microbiota at the 'omics' age. Insect Sci. 2010;**17**:199-219

[7] Nair RR, Vasse M, Wiegloss S, Sun L, Nicco YT, Velicer GJ. Bacterial predator-prey coevolution selects on virulence-associated prey defenses. bioRxiv. 2018;**398495**:1-34. DOI: 10.1101/398495

[8] Grabher MG, Haas BJ, Yassour M, Levin JZ, Thompson DA, Amit I, et al. Trinity: Reconstructing a full-length transcriptome without a genome from RNA-Seq data. Nature Biotechnology. 2011;**29**(7):644-652

[9] Jackson CR, Liew KC, Yule KM. Strucutral and functional changes in depht in microbial communities in a Tropical Malaysian Peat Swamp Forest. Microbial Ecology. 2009;**57**:402-412

[10] Kanokatrana P, Uengwetwanit T, Rattanachomsri U, Bunterngsook B, Nimchua T, Tangphatsornruang S, et al. Insights into the phylogeny and metabolic potential of a primary tropical peat swamp forest microbial community by metagenomic analysis. Microbial Ecology. 2011;**61**:518-528

[11] Jan C, Petersen JM, Werner J, Teeling H, Huan S, Glöckner FO, et al. The gill chamber epibiosis of deep-sea shrimp *Rimicaris exoculata*: An in-depth metagenomic investigation and discovery of *Zetaproteobacteria*. Environmental Microbiology. 2014;**16**(9):2723-2738

[12] Estrada-Peña A, Sprong H, Cabezas-Cruz A, De La Fuente J, Ramo A, Coipan EC. Nested coevolutionary networks shape the ecological relationships of ticks, hosts, and the Lyme disease bacteria of the Borrelia burgdorferi (s.l.) complex. Parasites & Vectors 2016;**9**(517): 1-15. DOI 10.1186/s13071-016-1803-z

[13] Kurtz J, Schulenburg H, Reusch TBH. Host–parasite coevolution—rapid reciprocal adaptation and its genetic basis. Zoology. 2016;**119**:241-243

[14] Rojas-Herrera R, Narváez-Zapata J, Zamudio-Maya M, Mena-Martínez ME. A Simple Silica-based method for metagenomic DNA extraction from soil and sediments. Molecular Biotechnology. 2008;**40**(1):13-17

[15] Chomzcsinsky P, Sacchi N. Single-step method for RNA isolation by acid guanidinium thyocyanate-phenol-chloroform extraction. Annals of Clinical Biochemistry. 1985;**162**(1): 156-159. DOI: 10.1006/abio.1987.9999

[16] Frias-López J, Shi Y, Tyson GW, Coleman ML, Schuster SC, Crisholm SW, et al. Microbial community gene expression in ocean surface waters. Proceedings of the National Academy of Sciences of the United States of America. 2010;**105**(10):3805-3810

[17] Valle-Gough RE, Apodaca-Hernández JE, Rodriguez-Dorantes M, Arena-Ortiz ML. Metatrancriptomic analysis from the Hepatopancreas of adult white leg shrimp (Litopenaeus vannamei). Symbiosis. 2018;**76**(1):51-62. DOI: 10.1007/s13199-017-0534-z

[18] Xiong J, Dai W, Qiu Q, Zu J, Yang W, Li C. Response of host–bacterial colonization in shrimp to developmental stage, environment and disease. Molecular Ecology. 2017;**27**(18): 3686-3699. DOI: 10.1111/mec.14822

[19] Bashiardes S, Zilberman-Schapira G, Elinav E. Use of metatranscriptomics in microbiome research. Bioinformatics and Biology Insights. 2016;**10**:19-25. DOI: 10.4137/BBI.S34610

[20] Feng F, Chen M, Xu C, Bai L, Yin X, Li X, et al. Transcriptomic analysis of Chinese bayberry (Myrica rubra) fruit development and ripening using RNA-Seq. BMC Genomics. 2012;**13**:19

[21] Frías-López J, Durán-Pinedo A. Effect of periodontal pathogens on the metatranscriptome of a healthy multispecies biofilm model. Journal of Bacteriology. 2012;**194**(8):2082-2095. DOI: 10.1128/JB.06328-11

[22] Komobis E, Cabellos L, Aguilar F, Frias-López C, Rpzas J, Marco ZR. TRUFA: A user-friendly web server for *de novo* RNA-seq analysis using cluster computing. Evolutionary Bioinformatics. 2015;**11**:97-104. DOI: 10.4137/EBO.S23873

[23] Schmieder R, Edwards E. Quality control and preprocessing of metagenomic datasets. Bioinformatics. 2011;**27**(6):863-864

[24] Keegan KP, Glass EM, Meyer F. MG-RAST, a metagenomics service for analysis of microbial community structure and function. Methods in Molecular Biology. 2016;**1399**:207-233. DOI: 10.1007/978-1-4939-3369-3_13

[25] Babendreier D, Joller D, Romeis J, Bigler F, Widmer F. Bacterial community structures in honeybee intestines and their response to two insecticidal protein. FEMS Microbiology Ecology. 2007;**59**(3):600-610. DOI: 10.1111/j.1574-6941.2006.00249.x

[26] McFall-Ngai M, Hadfieldb MG, Thomas I, Boschc CG, Hannah VC, Domazet-Loso T, et al. Animals in a bacterial world, a new imperative for the life sciences. PNAS. 2013;**110**(9):3229-3236. DOI: 10.1073/pnas.1218525110

[27] Papkou A, Gokhale CS, Schulenburg H. Host–parasite coevolution: Why changing population size matters. Zoology. 2016;**119**:330-338

[28] Hamdi C, Balloi A, Essanaa CE, Gonella E, Raddadi N, Ricci I, et al. Gut microbiome dysbiosis and honey bee heatlth. Journal of Applied Entomology. 2011;**135**(7):524-533. DOI: 10.1111/j.1439-0418.2010.01609.x

[29] Schloss PD, Wescott SL, Ryabin T, Hall JR, Hartmann M, Hollister EB, et al. Introducing mothur: Open-source, platform-independent, community-supported software for describing and comparing microbial communities. Applied and Environmental Microbiology. 2009;**75**(23):7537-7541. DOI: 10.1128/AEM.01541-09

[30] Bolyen E, Rideout JR, Dillon MR, Bokulich NA, Abnet C, Al-Ghalith GA, et al. QIIME 2: Reproducible, interactive, scalable, and extensible microbiome data science. PeerJ.

2018;**6**:e27295v2. DOI: 10.7287/peerj.preprints.27295v2

[31] Caporaso JG, Bittinger K, Bushman FD, DeSantis TZ, Andersen GL, Knight R. PyNAST: A flexible tool for aligning sequences to a template alignment. Bioinformatics. 2010;**26**(2):266-267

[32] Caporaso JG, Kuczynski J, Stombaugh J, Bittinger K, Bushman FD, Costello EK, et al. QIIME allows analysis of high-throughput community sequencing data. Nature Methods. 2010;7:335-336. DOI: 10.1038/nmeth.f.303

[33] Edgar RC. Search and clustering orders of magnitude faster than BLAST. Bioinformatics. 2010;**26**(19):2460-2461. DOI: https://doi.org/10.1093/bioinformatics/btq461

[34] Afgan E, Baker D, Batut B, van den Beek M, Bouvier D, Čech M, et al. The Galaxy platform for accessible, reproducible and collaborative biomedical analyses. Nucleic Acids Research. 2018;**46**(1):537-W544. DOI: 10.1093/nar/gky379

[35] Bengtsson-Palme J, Boulund F, Fick J, Kristiansson E, Larsson DGJ. Shotgun metagenomics reveals a wide array of antibiotic resistance genes and mobile elements in a polluted lake in India. Frontiers in Microbiology. 2014;**5**(648):1-14. DOI: 10.3389/fmicb.2014.00648

[36] Wood DE, Salzberg SL. Kraken: Ultrafast metagenomic sequence classification using exact alignments. Genome Biology. 2014;**15**:R46

[37] Conesa A, Götz S, García-Gómez JM, Terol J, Talón M, Robles M. Blast2GO: A universal tool for annotation, visualization and analysis in functional genomics research. Bioinformatics. 2005;**21**(18):3674-3676. DOI: 10.1093/bioinformatics/bti610

[38] Engel P, Moran N. The gut microbiota of insects – diversity in structure and function. FEMS Microbiology Reviews. 2013;**37**(5):699-735. DOI: 10.1111/1574-6976.12025

[39] Cornejo-Granados F, Gallardo-Becerra L, Leonardo-Reza M, Ochoa-Romo J, Ochoa-Leyva A. A meta-analysis reveals the enviromental and host factors shaping the structure and function of shrimp microbiota. PeerJ. 2018;**6**:e5682

[40] Power P, Berlemont R. Antarctic Soil Metagenome. In: Highlander Sk, Rodriguez-Valera F, White BA (eds) Encyclopedia of Metagenomics. Boston, MA: Springer; 2015. pp. 1-7. DOI: https://doi.org/10.1007/978-1-4899-7475-4

[41] Truu M, Juhansson J, Truu J. Microbial biomass, actitvity and community in constructed wetlands. Science of the Total Environment. 2009;**407**:3958-3971

[42] Cui J, Xiao M, Liu M, Wang Z, Liu F, Guo L, et al. Coupling metagenomics with cultivation to select host-specific probiotic micro-organisms for subtropical aquaculture. Journal of Applied Microbiology. 2017;**123**:1274-1285. DOI: 10.1111/jam.13555

[43] Li E, Xu C, Wang X, Wang S, Zhao Q, Zhang M, et al. Gut microbiota and its modulation for healthy farming of pacific white shrimp *Litopenaeus vannamei*. Reviews in Fisheries Science & Aquaculture. 2018;**26**(3):381-399. DOI: 10.1080/23308249.2018.1440530

[44] Porchas-Cornejo MA, Martínez-Porchas M, Vargas-Albores F, Gollas-Galvan T, Martínez-Córdova LR, Vazquez-Euan R, et al. High-resolution detection of bacterial profile of ocean water, before and after being used by shrimp farms. Aquaculture International. 2017;**25**(5):1833-1843

[45] Tzeng TD, Pao YY, Chen PC, Weng FCH, Jean WD, Wang D. Effects of host phylogeny and habitats on gut microbiomes of oriental river prawn (Macrobrachium nipponense). PLoS One. 2015;**10**(7):e0132860

[46] Vargas-Albores F, Porchas-Cornejo MA, Martínez-Porchas M, Villalpando-Canchola E, Gollas-Galván T, Martínez-Córdova LR. Bacterial biota of shrimp intestine is significantly modified by the use of a probiotic mixture: A high throughput sequencing approach. Helgoland Marine Research. 2017;**71**:5

[47] Crotti E, Baloi A, Hamdi C, Sansono L, Marzoarati M, Gonella E, et al. Microbial symbionts: A resource for the management of insect related problems. Microbial Biotechnology. 2012;**5**(3):307-317

[48] Mohr KI, Tebbe CC. Diversity and phylotype consistency of bacteria in the guts of three bee species (*Apoidea*) at an oilseed rape field. Environmental Microbiology. 2006;**8**(2):258-272. DOI: 10.1111/j.1462-2920.2005.00893.x

[49] Macfarlane S, Dillon JF. Microbial biofilms in the human gastrointestinal tract. Journal of Applied Microbiology. 2007;**102**(5):1187-1196. DOI: 10.1111/j.1365-2672.2007.03287.x

[50] Turnbaugh PJ, Hamady M, Yatsunenko T, Cantarel BL, Duncan A, Ley RE, et al. A core gut microbiome in obese and lean twins. Nature. 2008;**457**(7228):480-484. DOI: 10.1038/nature07540

[51] Shen Z, Zhang N, Mustapha A, Xu D, Deng D, Reed M, et al. Identification of host-specific genetic markers within 16S rDNA intervening sequences of 743 genera of fecal bacteria. Journal of Data Mining in Genomics and Proteomics. 2016;7(1):1-9

[52] Purroy N, Wu CJ. Co-evolution of leukemia and host immune cells in chronic lymphocytic leukemia (CLL). Cold Spring Harbor Perspectives in Medicine. 2018;**7**(4):1-28. DOI: 10.1101/cshperspect.a026740

[53] Langille MGI, Zaneveld J, Caporaso JG, McDonald D, Knights D, Reyes J, et al. Predictive functional profiling of microbial communities using 16S rRNA marker gene sequences. Nature Biotechnology. 2013;**1**(10):8

[54] Parks DH, Tyson GW, Hugenholtz P, Beiko RG. STAMP: Statistical analysis of taxonomic and functional profiles. Bioinformatics. 2014;**30**:3123-3124

[55] Koo H, Mojib N, Hakim JA, Hawes I, Tanabe Y, Andersen DT, et al. Microbial communities and their predicted metabolic functions in growth laminae of a unique large conical mat from Lake Untersee, East Antarctica. Frontiers in Microbiology. 2017;**8**:1347. DOI: 10.3389/fmicb.2017.01347

[56] Ezenwa VO, Gerardo NM, Inouye DW, Medina M, Xavier JB. Animal Behavior and Microbiome. Science. 2012;**338**(6104):198-199. DOI: 10.1126/science.1227412

[57] Langmead B, Trapnell C, Pop M, Salzberg SL. Ultrafast and memory-efficient alignment of short DNA sequences to the human genome. Genome Biology. 2009;**10**(3):25.1-25.10

[58] Magoč T, Salzberg SL. FLASH: Fast length adjustment of short reads to improve genome assemblies. Bioinformatics. 2011;**1**(21):2957-2963. DOI: 10.1093/bioinformatics/btr507

[59] Namiki T, Hachiya T, Tanaka H, Sakakibara Y. MetaVelvet: An extension of Velvet assembler to de novo metagenome assembly from short sequence reads. Nucleic Acids Research. 2012;**40**(20):e155

[60] Treangen TJ, Koren S, Sommer DD, Liu B, Astrovskaya I, Ondov B, et al. MetAMOS: A modular and open source metagenomic assembly and analysis pipeline. Genome Biology. 2013;**14**(1):R2. DOI: 10.1186/gb-2013-14-1-r2

[61] Xie Y, Wu G, Tang J, Luo R, Patterson J, Liu S, et al. SOAPdenovo-Trans: De novo transcriptome assembly with short RNA-Seq reads. Bioinformatics. 2014;**30**(12):1660-1666

[62] Zerbino DR, Birney E. Velvet: Algorithms for de novo short read assembly using de Bruijn graphs. Genome Research. 2008;**18**(5):821-829. DOI: 10.1101/gr.074492.107

[63] Zhang J, Kobert K, Flouri T, Stamatakis A. PEAR: A fast and accurate Illumina Paired-End reAd mergeR. Bioinformatics. 2014;**30**(5):614-620

[64] Thureborn P, Franzetti A, Lundin D, Sjölin S. Reconstructing ecosystem functions of the active microbial community of the Baltic Sea oxygen depleted sediments. PeerJ. 2016;**4**:e1593. DOI: 10.7717/peerj.1593

[65] Darzi Y, Letunic I, Bork P, Yamada T. iPath3.0: Interactive pathways explorer v3. Nucleic Acids Research. 2018;**46**(1):510-513

Chapter 5

Metagenomics-Based Phylogeny and Phylogenomic

Ayixon Sánchez-Reyes and Jorge Luis Folch-Mallol

Abstract

Phylogenetic relationships among microbial taxa in natural environments provide key insights into the mechanisms that shape community structure and functions. In this chapter, we address the current methodologies to carry out community structure profiling, using single-copy markers and the small subunit of the rRNA gene to measure phylogenetic diversity from next-generation sequencing data. Furthermore, the huge amount of data from metagenomics studies across the world has allowed us to assemble thousands of draft genomes, making necessary the comparison of whole genomes composites through phylogenomic approximations. Several computational tools are available to carry out these analyses with considerable success; we present a compendium of those open source tools, easy to use and with modest hardware requirements, with the aim that they can be applied by biologists non-specialists to study microbial diversity in a phylogenetic context.

Keywords: metagenomics profiling, phylogenetic diversity, phylogenetic metadata representation

1. Introduction

Next-generation sequencing technologies have transformed our perception of diversity and microbial distribution in natural ecosystems and have contributed substantially to the discovery of totally new microbial landscapes in such distinctive environments as the gut of mammals, the vegetal rhizosphere, vascular tissues of higher plants, and even in volcanic lakes [1–3]. There are two general approaches to profile microbial communities through next-generation sequencing techniques: shotgun sequencing of total DNA isolated directly from the environment and sequencing of variable regions coming from SSU-rRNA genes (we know these approaches as metagenomics methods since all involve the culture-independent genomic analysis of microbiomes on a particular environment [4, 5]). Both approaches have been widely used to trace microbial diversity at increasingly fine taxonomic levels, either by capturing a representative fraction of the total gene content or by amplicon sequencing techniques like the popular bacterial 16S rRNA. Each method has advantages and disadvantages, and the selection depends on several factors like taxonomic level resolution, cost, sensitivity, and primer bias, among others. One of the challenges associated with metagenomics methods is the analysis of massively generated data. Both the sequencing of amplicons and environmental DNA produces millions of short DNA sequences (reads), which must undergo preprocessing and quality control, before they can be used to extract biologically useful information from them. One of the goals of massively sequencing data analysis is to obtain the patterns of phylogenetic diversity in ecological communities, an important trait in order to assess the classic ecological questions "Who is there?" or "What they are doing?" and provide better understandings into the phylogenetic relationships among microbial community taxa. Extracting phylogenetic information from massive sequencing reads is not a trivial task; however, it can be achieved with reasonable success by using several profiling tools adapted both to the analysis of amplicons of ribosomal genes and to the conserved genes between different domains [6, 7]. The microbial community structure has been approached mostly using the 16S SSU-rRNA gene as phylogenetic marker, mainly due to lower sequencing costs and an acceptable relation of specificity-resolution in taxonomic assignments [8], while methods that use single-copy markers obtained from shotgun sequencing reads or assembled samples are gaining relevance because they have demonstrated strain-level resolution [9, 10], a really hard issue when analyzing complex microbiomes.

To date, several computational tools have been developed to carry out community profiling and phylogenetic inferences from next-generation sequencing data with considerable success. In this chapter we present a compendium of open-source tools and easy-to-use with modest hardware requirements, with the aim that they can be applied by biology non-specialists to study microbial diversity in a phylogenetic context. We show several practical examples explained step by step, in order to provide to the reader, the replication using their own data.

We have selected tools for use on a local computer through the Unix command line, and tools are available from dedicated servers, with easy access and intuitive use. The examples described in the chapter were tested on a Dell Optiplex 7010 desktop, 6T6ZYV1 Series, Intel (R) Core (TM) i5-3550 CPU at 3.30 GHz, Memory 12 GiB.

2. Community structure profiling across microbial samples using single-copy markers

With the advent of massive DNA sequencing technologies, several methods have been developed to assign shotgun reads to microbial taxonomic categories. These methods aim to perform a microbial community profiling that infers its relative structure, and they are very important to understand how microbiomes work in nature, their phylogenetic composition, and even their dynamics and evolutionary history. The starting point for these analyzes is a set of reads obtained by massive sequencing whose length is variable (as little as 50–75 bp up to >1000 bp) depending on the platform used (Illumina, Ion Torrent, PacBio RS). We can understand by a read the sequence of bases from a single discrete molecule of DNA, obtained in a massively parallel manner [11]. However, currently most metagenomics studies use a range of a short-read sequencing instruments between 100 and 600 bp in order to maximize counting reads and lower costs. These short-reads contain the genomic, phylogenetic, and functional information of the microbiome into millions of discrete DNA fragments, which are sufficient to make a reliable estimate of the phylogenetic diversity present in a microbial sample (**Figure 1**).

The taxonomic composition of a microbial community can be estimated from a set of short-reads by assigning each read to the most likely microbial lineage [12]. Historically, a single gene target approach has been the gold standard for assigning taxonomy in the Prokaryote domain, through the 16S ribosomal RNA gene. However, this presents important biases related to copy-number variations and significant intraspecific differences ~6%. In this sense, both clade-specific and universal single-copy phylogenetic markers genes have gained popularity among the

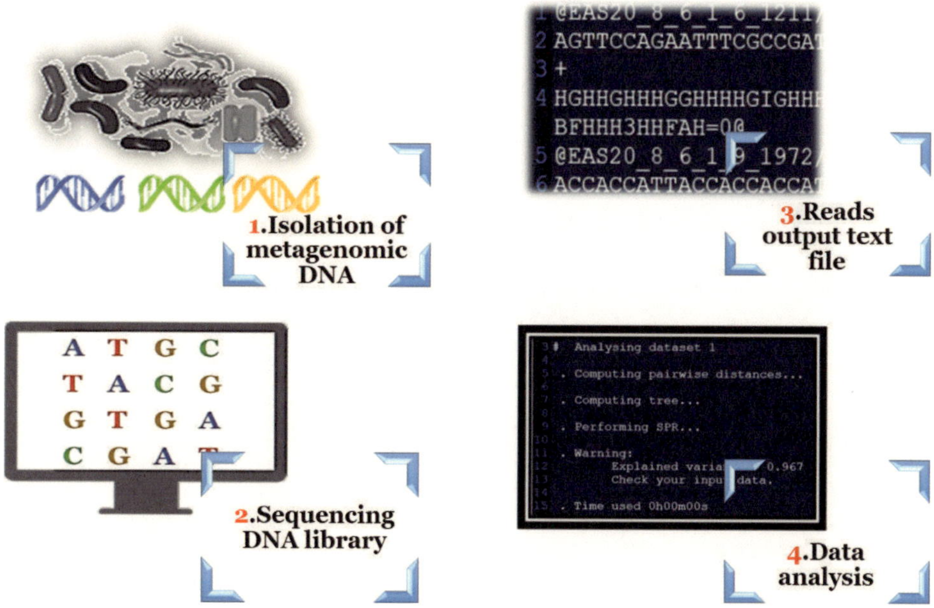

Figure 1.
General overview of metagenomics analysis in a microbial sample by next-generation sequencing: (1) isolation of metagenomic DNA, (2) sequencing DNA library, (3) reads output text file, and (4) data analysis.

scientific community since they are not subject to intragenomic diversity, are rarely subjects of horizontal transfer, and have proven robustness to delineate species and prokaryotic strains in multiple studies, because several genes can be combined to reconstruct phylogenies [13, 14]. Although each method selects its own set of clade-specific or universal markers, most of these genes encode proteins with functional relevance in housekeeping metabolism (**Table 1**). To make the analysis, the coding nucleotide sequences are generally used as they offer better resolution than amino acid sequences in closely related organisms [16]. This simplifies the computational analysis as the short-reads could be compared unambiguously without the need to translate them into proteins, which could generate artifacts given the small size of the reads.

One of the most popular tools for microbial profiling based on clade-specific marker genes is the MetaPhlAn classifier [12, 17]. MetaPhlAn maps the experimental reads against a collection of 231 markers for species-level comparisons and >115,000 markers for higher taxonomic levels. Among the advantages of this classifier is that no preprocessing is required, so raw data can be uploaded and analyzed. The main disadvantage for non-specialists is that MetaPhlAn works through the command line in a Unix architecture.

2.1 Profiling a textile dye degrader microbiome with MetaPhlAn2

Next we described the steps performed for profiling a microbial community capable of degrading the textile dye HC Blue no. 2. Also we show a graphical representation of the profiling phylogenetic metadata. This general strategy can be applied to profile any microbial community from short-reads obtained by massive sequencing. Symbol convention: Comments (#); executable commands ($). The raw data are available on [18].

You can find a complete MetaPhlAn guide on the author's site: https://bitbucket.org/biobakery/biobakery/wiki/metaphlan2.

Clusters of orthologous groups of proteins (COG)	Protein name
COG0048	Ribosomal protein S12
COG0049	Ribosomal protein S7
COG0052	Ribosomal protein S2
COG0080	Ribosomal protein L11
COG0081	Ribosomal protein L1
COG0085	DNA-directed RNA polymerase, beta subunit
COG0087	Ribosomal protein L3
COG0088	Ribosomal protein L4
COG0090	Ribosomal protein L2
COG0091	Ribosomal protein L22
COG0092	Ribosomal protein S3
COG0093	Ribosomal protein L14
COG0094	Ribosomal protein L5
COG0096	Ribosomal protein S8
COG0097	Ribosomal protein L6P/L9E
COG0098	Ribosomal protein S5
COG0099	Ribosomal protein S13
COG0100	Ribosomal protein S11
COG0102	Ribosomal protein L13
COG0103	Ribosomal protein S9
COG0124	Histidyl-tRNA synthetase
COG0172	Seryl-tRNA synthetase
COG0184	Ribosomal protein S15P/S13E
COG0185	Ribosomal protein S19
COG0186	Ribosomal protein S17
COG0197	Ribosomal protein L16/L10E
COG0200	Ribosomal protein L15
COG0201	Preprotein translocase subunit SecY
COG0202	DNA-directed RNA polymerase, alpha subunit/40 kD subunit
COG0215	Cysteinyl-tRNA synthetase

Table 1.
Universal single-copy phylogenetic marker genes employed in metagenomics-based phylogenies for delineation of prokaryotic species (modified from [15]).

\# **Installing MetaPhlAn2**
\# with an activated Bioconda channel in Linux, type the following command:
$ *conda install metaphlan2*
\# this will install the software with all its dependencies
\# **Generate a taxonomic profile**
\# Type the following command:
$ *python /path/to/metaphlan2.py /path/to/textile_microbiome.fastq.gz --input_type fastq > textile_microbiome _profile.txt*

#Sample ID	Abundance MetaPHlAn2_analysis (%)
k__Bacteria	56.02708
k__Archaea	43.94783
k__Viruses	0.02509
k__Bacteria|p__Firmicutes	45.48396
k__Archaea|p__Euryarchaeota	43.94783
k__Bacteria|p__Proteobacteria	8.46518
k__Bacteria|p__Actinobacteria	2.07794
k__Viruses|p__Viruses_noname	0.02509

The taxonomic levels are: Kingdom, k; and Phylum, p. The table was trimmed to show only up to the phylum level; to read complete report, see [7].

Table 2.
Features of the abundance table for a textile dye degrader microbiome profile.

The output profile (called: textile_microbiome _profile.txt) contains the computed clade's abundances (**Table 2**).
 # Capture phylogenetic relatedness with GraPhlAn
In order to visualize microbial abundances on a phylogeny we'll use GraPhlAn tool [19].
Installing GraPhlAn
Type the following two commands:
$ brew tap biobakery/biobakery
$ brew install graphlan
# In order to know the installation directory type the following command:
$ which graphlan
 # Input files
Type the following commands sequentially:
*$ python path/to/merge_metaphlan_tables.py *_profile.txt > merged_abundance_table.txt*
$ python path/to/export2graphlan.py --skip_rows 1,2 -i merged_abundance_table.txt --tree merged_abundance.tree.txt --annotation merged_abundance.annot.txt --most_abundant 100 --abundance_threshold 1 --least_biomarkers 10 --annotations 5,6 --external_annotations 7 --min_clade_size 1
 # Create the phylogeny
Type the following commands sequentially:
$ python path/to/graphlan_annotate.py --annot merged_abundance.annot.txt merged_abundance.tree.txt merged_abundance.xml
$ python path/to/graphlan.py --dpi 300 merged_abundance.xml merged_abundance.png --external_legends

Finally, you will obtain:

- a cladogram called: merged_abundance.png
- an annotation file called: merged_abundance_annot.png
- a legend file called: merged_abundance_legend.png

You can change the format of the final results to pdf, just modifying the name: *merged_abundance.png* to *merged_abundance.pdf* in the last command. A representation of the annotated cladogram is shown in **Figure 2**. The size of the nodes correlates with microbial community relative abundances.

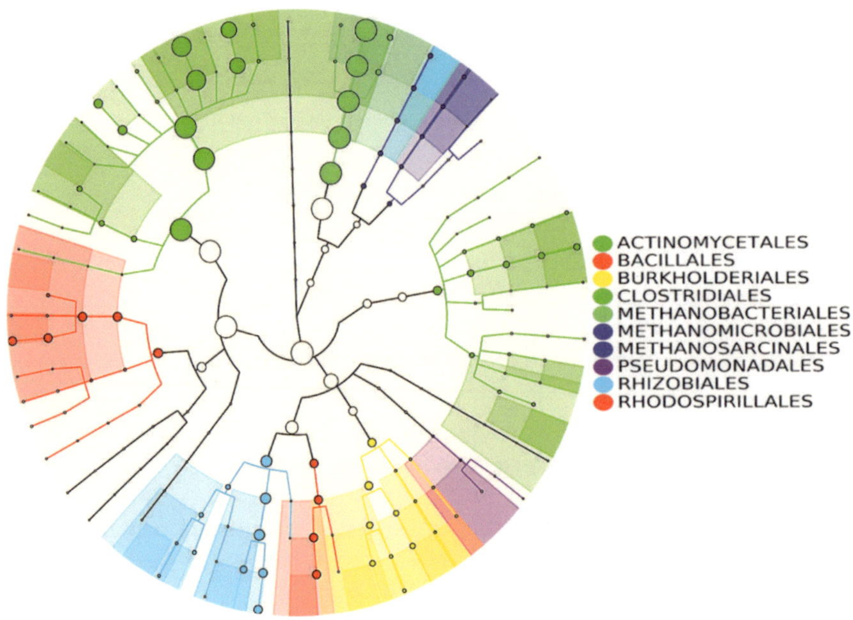

Figure 2.
Cladogram produced by GraPhlAn software with metadata from MetaPhlan community profiling to order level. Taxon abundance is proportional to the circle diameter.

3. Phylogenetic diversity of microbial communities based on 16S rDNA gen

Estimating the taxonomic and phylogenetic diversity of a microbial community is also possible through sequencing and analysis of small ribosomal RNA subunit (16S rRNA) gene, whenever this sequence has been considered for a long time a stable marker, crucial in the microbial systematics of the last 30 years. 16S ribosomal ribonucleic acid is a key component of the small subunit of prokaryotic ribosomes, central player in the cellular biology of microorganism; it serves as a linker for the process of translating genetic information to proteins [20]. Because DNA is much easier to sequence than RNA, DNA segment coding for 16 rRNA is obtained for the purposes of sequencing (**Figure 3**). This gene fragment meets several features that have made it a "quasi-gold standard" for bacterial taxonomy:

- It is a ubiquitous gene in the Bacteria and Archaea domains.
- Within its ~1500 bp, it has discrete regions with enough variability to establish a phylogenetic signal among phyla and even genus.
- It has conserved regions that allow the design of "universal primers," a very useful feature in massive sequencing.
- It has several databases enriched with sequences from almost all international projects where 16S sequences are obtained (**Table 3**). For example, the 16S ribosomal RNA (Bacteria and Archaea) database from the National Center for Biotechnology Information (NCBI) contains near to 20,831 curated records and more than 17 million of total records (consulted date: 2019/08/05: https://blast.ncbi.nlm.nih.gov/Blast.cgi?PROGRAM=blastn&PAGE_TYPE=BlastSearch&LINK_LOC=blasthome).

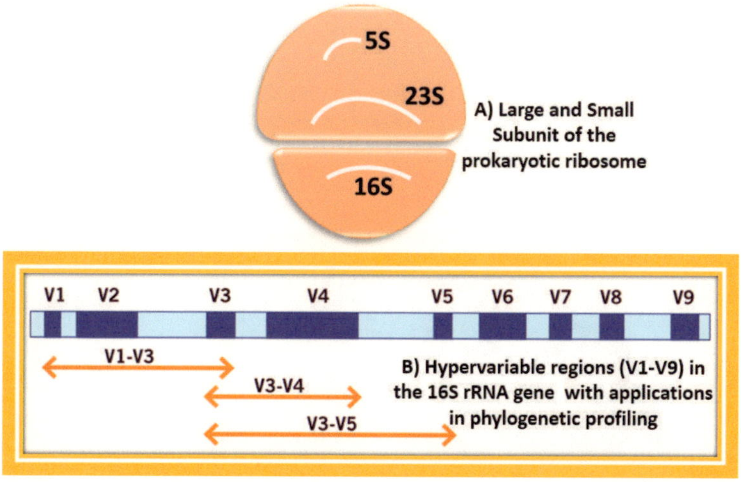

Figure 3.
Prokaryotic ribosome general representation and variable sequence regions used in microbial phylogenetic diversity estimations.

Database	Available SSU sequences	Current release	Citation/link
16S NCBI database	20831[a]	2019	[21]/https://blast.ncbi.nlm.nih.gov/
SILVA rRNA database	23629[b]	2018	[22]/https://www.arb-silva.de
Ribosomal Database Project (RDP)	16277	September 2016	[23]/https://rdp.cme.msu.edu
EzBioCloud 16S database	13132[c]	2019	[24]/https://www.ezbiocloud.net/
Genomic-based 16S ribosomal RNA database (GRD)	13202	2015	https://metasystems.riken.jp/grd/download.html

[a]*Not redundant manually curated small (16S, SSU) subunit ribosomal RNA sequences.*
[b]*The dataset contains 23,629 SSU sequences representing a single bacterial type strain up to June 2017.*
[c]*Phylotypes with validly published names.*

Table 3.
Most popular public databases for depositing and analyzing sequences of the 16S ribosomal gene.

3.1 16S community profiling by analysis of ribosomal amplicons

Microbial diversity is measured as a function that depends on the richness and abundance of distinct taxons among any community [25]. Obtaining representative DNA sequences from the entire community is essential to make valid inferences. Profiling a microbial community through 16S gene analysis generally consists of four steps (**Figure 4**). To date, several computational tools have been developed to analyze microbial communities through the 16S gene marker; however, estimating the total microbial diversity in any environment is a still a major challenge [6, 26–28], influenced by several factors, among them we want to mention two: (I) processing huge amounts of data moves within the limits of modern computing and (II) the need for some expertise that can cost years of training. Fortunately, many tools have been developed in recent years, aiming to make bioinformatics platforms dedicated to this type of analysis more human-friendly, and there are dedicated sites exclusively to deposit computational alternatives for almost all needs, for example, https://github.com/.

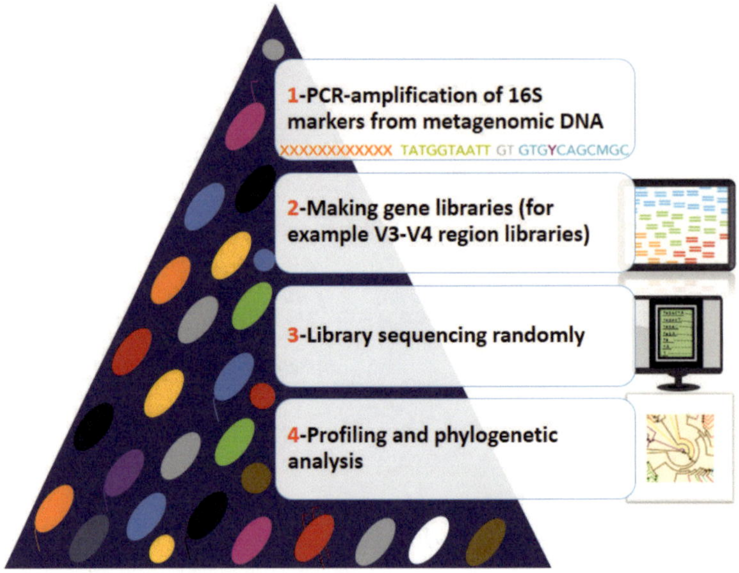

Figure 4.
General steps for profiling a microbial community through 16S gene analysis.

A good example of these multiplatforms to profile microbial communities is the Microbiome Taxonomic Profiling (MTP) pipeline from EzBioCloud site (https://www.ezbiocloud.net/contents/16smtp) [24]. Among its fundamental advantages are: it is free, knowledge of Linux environment is not needed to carry out the analyses, and several types of outputs such as functional profiles, taxonomic and phylogenetic structure, as well as on-demand comparison with other published microbiome data are fully available. New users of EzBioCloud will be required to open a local server account (https://www.ezbiocloud.net/signup?from=addMTP); after that you can upload up to 100,000 reads for sample and begin the analysis. We list general steps to perform a profiling on the platform (**Box 1**).

The platform consists of a very intuitive and user-friendly presentation that guides the beginner user at every stage of the analysis. The first step is the uploading of the next-generation sequencing data (16S amplicon reads). After that, you can request for the MTP pipeline, and the analysis starts. In a relatively short time, you can access the result portal with the preprocessing results resumed in pre-filtered reads (by removing low-quality and chimeric amplicons), statistics about read lengths, and taxonomic read assignments at species level.

Other outputs in results portal are related with several diversity indices, taxonomic composition and hierarchy, and graphical implementations like Krona [29]. MTP implements seven different diversity indices; among them is the phylogenetic diversity index, a measure of biodiversity that considers phylogenetic difference between taxons and ponders several variables like taxonomic diversity and species abundances or distributions.

3.2 Extracting 16S sequences from assembled data

In occasions, we do not have a set of DNA short-reads, but assembled composites in contiguous regions of variable size. Such is the case of genomes assembled from metagenomes or contigs from complex metagenomes. Inferring taxonomic diversity from this type of data usually requires other strategies. One of the most useful is to predict all the rRNA sequences contained in the assembly and cluster them

DOI: http://dx.doi.org/10.5772/intechopen.89492

EZ BioCloud

BOX 1. 16S -based microbiome taxonomic profiling pipeline used in EzBioCloud

1-Raw data type:
- Single-end reads (Roche 454)
- Single or paired-end reads (Illumina MiSeq, HiSeq, others)
- CCS (circular consensus sequencing) reads (Pacific Biosciences (PacBio))
- Other NGS data in FASTQ or FASTA format outputs

2-Preprocessing:
- Merging paired-end reads
- Trimming primers
- Filtering by quality
- Denoising and non-redundant reads trimming

3-Taxonomic and phylogenetic profiling:
- Taxonomic assignment
- Clustering sequences in operational taxonomic units (OTUs)
- Estimating alpha diversity indices
- Secondary analysis using EzBioCloud 16S-based pipeline

Box 1.
16S-based microbiome taxonomic profiling pipeline used in EzBioCloud.

according to their identity (this implies making a list of nonredundant sequences) to define operative taxonomic units. A simple way to address this problem is through the use of Barrnap software [27]; it works through the Unix command line and has the advantage of consuming few computational resources, so that several complex microbiomes can be analyzed in a personal computer for extraction of rRNA sequences. Barrnap gives us an output with all predicted sequences; this includes 5S, 16S, and 23S rRNA in the case of bacteria. The sequences can be saved on-demand in a text file and subsequently analyzed by a third-party phylogenetic processing software to establish evolutionary relationships between taxa. A suitable platform for this objective is SeaView [28], which contains sequence alignment and curing utilities, as well as a set of phylogenetic reconstruction methods, like PhyML, which uses maximum likelihood algorithms and seven different evolutionary models. It is also possible to use distance methods such as Neighbor Joining and BioNeighbor Joining, both with seven different methods to calculate distances between sequences. The platform is open access and has the advantage of being a graphical application that works on Unix and Windows, as well as being very intuitive.

4. Open-source software for phylogenetic and phylogenomic surveys

Genome-based comparisons play an essential role in the current taxonomy and phylogenetic of Bacteria and Archaea domains and eventually will replace the single gene target approach ruled by 16S rRNA gene phylogeny. The exponential growth of complete genomes and genome drafts with significant completeness values and low contamination (<5%) in international databases has resulted in an approach to phylogenetic analysis where the whole information has become in a more conservative

Software	Application	NGS data	License	Environment	Reference
MetaPhlAn	Microbial community profiling	Shotgun sequencing data	Open access	Unix command line	[17]
FOCUS	Taxonomic profiling	Shotgun unannotated sequencing reads	Open access	Unix command line and Web implementation	[30]
Kraken	Assigning taxonomic labels to metagenomic DNA sequences	Shotgun unannotated sequencing reads	Open access	Unix command line	[31]
GraPhlAn	Phylogenetic analysis	Metadata from short-read community profiling	Open access	Unix command line	[19]
PICRUSt	Predictive functional profiling of microbial communities	16S amplicons	Open access	Unix command line	[32]
QIIME	Taxonomic and phylogenetic profiling	16S amplicons	Open access	Unix command line and web implementation	[28]
Mothur	Taxonomic and phylogenetic profiling	16S rRNA gene sequences	Open access	Unix command line	[6]
UBCG	Phylogenomic tree reconstruction	Set of bacterial genomes	Open access	Unix command line	[33]
GToTree	A user-friendly workflow for phylogenomics	Set of bacterial genomes	Open access	Unix command line	[34]
PhylOTU	Identifies OTUs from rRNA sequence by phylogenetic profiles	PCR and shotgun sequenced SSU-rRNA markers	freely available	Unix command line	[35]
PhyloSift	Phylogenetic analysis of genomes and metagenomes	Metagenomic datasets generated by modern sequencing platforms	Freely available	Unix command line	[36]
VITCOMIC2	Phylogenetic representation based on 16S rRNA gene amplicons	16S amplicons	Freely	Web server	[37]
Barrnap	Very fast ribosomal RNA prediction	Assemblies from genomic or metagenomic data	Freely available	Unix command line	[38]
SeaView	Multiplatform for phylogenetic inferences	DNA or protein sequences	Freely available	Unix and Windows environments	[39]

Table 4.
Open-source software for metagenomics-based profiling and phylogenies.

DOI: http://dx.doi.org/10.5772/intechopen.89492

fingerprint of the taxonomic categories. The current challenges for science involve improving existing methods for data acquisition and processing, since comparative analysis, even among modest-sized microbial genomes, can be computationally expensive. Here we present a list of those open-source tools and easy-to-use and modest hardware requirements, with the aim that they can be applied by biologists to study microbial diversity in a phylogenetic context (**Table 4**).

5. Conclusions

Profiling microbial communities from massive sequencing data constitutes a breaking point in the understanding of population structure and dynamics, their ecological functions and the complex relationships established between non-cultivable microorganisms. Through technological developments such as next-generation sequencing and the developing of hundreds of open-access platforms, we have been able to better understand the role of the microbial world in natural ecosystems. This chapter intends to bring the use of computational biology tools to professionals in biological sciences with different expertise, interested in the world of metagenomics analysis. We have started with the basics of microbial community profiling through shotgun sequencing data and its processing using MetaPhlAn software (the reader will notice that there are other tools perhaps more appropriate to their conditions, an interesting option is the FOCUS software that works through a Web server). MetaPhlAn has the advantage of being fully integrated with the GraPhlan phylogenetic reconstruction tools. We dedicate a complete section to the 16S gene-based communities profiling; we illustrate the EzBioCloud platform, a useful tool to obtain ecological and phylogenetic information of microbiomes. An alternative approach to process assembled data is the use of Barrnap software, which is very fast and efficient to extract ribosomal sequences in assembled data, which can be subsequently clustered and processed with phylogenetic construction tools such as SeaView. Finally, we present a list of software that can serve as a guide for the analysis of microbiomes from their taxonomic characterization to the study of phylogenetic relationships between taxa.

Acknowledgements

We thank the supercomputing resources and services of the Dirección General de Cómputo y de Tecnologías de Información y Comunicación (DGTIC) from Universidad Nacional Autónoma de México, through the project: LANCAD-UNAM-DGTIC-371.

Conflict of interest

The authors declare no conflict of interest.

Author details

Ayixon Sánchez-Reyes[1*] and Jorge Luis Folch-Mallol[2]

1 Cátedras Conacyt, Institute of Biotechnology-UNAM, Cuernavaca, Morelos, Mexico

2 Biotechnology Research Center, Universidad Autónoma del Estado de Morelos, Cuernavaca, Morelos, Mexico

*Address all correspondence to: ayixon.sanchez@mail.ibt.unam.mx

References

[1] Press MO, Wiser AH, Kronenberg ZN, Langford KW, Shakya M, Lo C-C, et al. bioRxiv [Internet]. 2017:198713. Cold Spring Harbor Laboratory. Available from: https://www.biorxiv.org/content/10.1101/198713v1 [Accessed: 08 October 2019]

[2] Bakker PAHM, Berendsen RL, Doornbos RF, Wintermans PCA, Pieterse CMJ. The rhizosphere revisited: Root microbiomics. Frontiers in Plant Science. 2013;**4**:1-7. DOI: 10.3389/fpls.2013.00165

[3] Mapelli F, Marasco R, Rolli E, Daffonchio D, Donachie S, Borin S. Microbial Life in Volcanic Lakes. In: Rouwet D, Christenson B, Tassi F, Vandemeulebrouck J, editors. Volcanic Lakes. Advances in Volcanology. Berlin, Heidelberg: Springer; 2015. p. 507-522. DOI: 10.1007/978-3-642-36833-2_23

[4] Handelsman J, Rondon MR, Brady SF, Clardy J, Goodman RM. Molecular biological access to the chemistry of unknown soil microbes: A new frontier for natural products. Chemistry & Biology. 1998;**5**:R245-R249. DOI: 10.1016/s1074-5521(98)90108-9

[5] Sleator R, Shortall C, Hill C. Metagenomics. Letters in Applied Microbiology. 2008;**47**:361-366. DOI: 10.1111/j.1472-765x.2008.02444.x

[6] Schloss PD, Westcott SL, Ryabin T, Hall JR, Hartmann M, Hollister EB, et al. Introducing mothur: Open-source, platform-independent, community-supported software for describing and comparing microbial communities. Applied and Environmental Microbiology. 2009;**75**:7537-7541. DOI: 10.1128/aem.01541-09

[7] Truong DT, Tett A, Pasolli E, Huttenhower C, Segata N. Metagenomic phylogenetic analysis. Nature Methods. 2017;**12**(10):626-638. Available from: http://www.nature.com/doifinder/10.1038/nmeth.3589

[8] Schriefer AE, Cliften PF, Hibberd MC, Sawyer C, Brown-Kennerly V, Burcea L, et al. A multi-amplicon 16S rRNA sequencing and analysis method for improved taxonomic profiling of bacterial communities. Journal of Microbiological Methods. 2018;**154**:6-13. DOI: 10.1016/j.mimet.2018.09.019

[9] Silva GGZ, Green KT, Dutilh BE, Edwards RA. SUPER-FOCUS: A tool for agile functional analysis of shotgun metagenomic data. Bioinformatics. 2015;**32**:354-361. DOI: 10.1093/bioinformatics/btv584

[10] Truong DT, Tett A, Pasolli E, Huttenhower C, Segata N. Microbial strain-level population structure and genetic diversity from metagenomes. Genome research. 2017;**27**:626-638. DOI: 10.1101/gr.216242.116

[11] Goodwin S, Mcpherson JD, Mccombie WR. Coming of age: Ten years of next-generation sequencing technologies. Nature Reviews Genetics. 2016;**17**:333-351. DOI: 10.1038/nrg.2016.49

[12] Segata N, Waldron L, Ballarini A, Narasimhan V, Jousson O, Huttenhower C. Metagenomic microbial community profiling using unique clade-specific marker genes. Nature Methods. 2012;**9**:811-814. DOI: 10.1038/nmeth.2066

[13] Sunagawa S, Mende DR, Zeller G, Izquierdo-Carrasco F, Berger SA, Kultima JR, et al. Metagenomic species profiling using universal phylogenetic marker genes. Nature Methods. 2013;**10**:1196-1199. DOI: 10.1038/nmeth.2693

[14] Mende DR, Sunagawa S, Zeller G, Bork P. Accurate and universal

delineation of prokaryotic species. Nature Methods. 2013;**10**(9):881-884. DOI: 10.1038/nmeth.2575

[15] Mende DR, Sunagawa S, Zeller G, Bork P. Accurate and universal delineation of prokaryotic species. Nature Methods. 2013;**10**:881-884. DOI: 10.1038/nmeth.2575

[16] Konstantinidis KT, Tiedje JM. Towards a genome-based taxonomy for prokaryotes. Journal of Bacteriology. 2005;**187**:6258-6264. DOI: 10.1128/jb.187.18.6258-6264.2005

[17] Truong DT, Franzosa EA, Tickle TL, Scholz M, Weingart G, Pasolli E, et al. MetaPhlAn2 for enhanced metagenomic taxonomic profiling. Nature Methods. 2015;**12**:902-903. DOI: 10.1038/nmeth.3589

[18] Sánchez-Reyes A. Massive sequencing dataset of a textile dye degrader microbiome [Internet]. Mendeley Data, v1. 2019. DOI: 10.17632/t4j8tc3njd.1

[19] Asnicar F, Weingart G, Tickle TL, Huttenhower C, Segata N. Compact graphical representation of phylogenetic data and metadata with GraPhlAn. PeerJ. 2015;**3**:1-17. DOI: 10.7717/peerj.1029

[20] Ramazzotti M, Bacci G. 16S rRNA-based taxonomy profiling in the metagenomics era. In: Metagenomics: Perspectives, Methods, and Applications. Cambridge, Massachusetts: Academic Press; 2017. pp. 103-119. DOI: 10.1016/B978-0-08-102268-9.00005-7

[21] Sayers EW, Agarwala R, Bolton EE, et al. Database resources of the national center for biotechnology information. Nucleic Acids Research. 2019;**47**:D23-D28. DOI: 10.1093/nar/gky1069

[22] Yilmaz P, Parfrey LW, Yarza P, Gerken J, Pruesse E, Quast C, et al. The SILVA and "all-species Living Tree Project (LTP)" taxonomic frameworks. Nucleic Acids Research. 2013;**42**:D643-D648. DOI: 10.1093/nar/gkt1209

[23] Maidak BL, Olsen GJ, Larsen N, Overbeek R, McCaughey MJ, Woese CR. The RDP (Ribosomal Database Project). Nucleic Acids Research. 1997;**25**:109-110. DOI: 10.1093/nar/25.1.109

[24] Yoon SH, Ha SM, Kwon S, Lim J, Kim Y, Seo H, et al. Introducing EzBioCloud: A taxonomically united database of 16S rRNA gene sequences and whole-genome assemblies. International Journal of Systematic and Evolutionary Microbiology. 2017;**67**:1613-1617. DOI: 10.1099/ijsem.0.001755

[25] Haegeman B, Hamelin J, Moriarty J, Neal P, Dushoff J, Weitz JS. Robust estimation of microbial diversity in theory and in practice. The ISME Journal. 2013;**7**(6):1092-1101

[26] Ramazzotti M, Berná L, Donati C, Cavalieri D. riboFrame: An improved method for microbial taxonomy profiling from non-targeted metagenomics. Frontiers in Genetics. 2015;**6**:1-12. DOI: 10.3389/fgene.2015.00329

[27] Bruno F, Marinella M, Santamaria M. e-DNA meta-barcoding: From NGS raw data to taxonomic profiling. Methods in Molecular Biology. 2015;**1269**:257-278. DOI: 10.1007/978-1-4939-2291-8_16

[28] Caporaso JG, Kuczynski J, Stombaugh J, Bittinger K, Bushman FD, Costello EK, et al. QIIME allows analysis of high-throughput community sequencing data. Nature Methods. 2010;**7**:335-336

[29] Ondov BD, Bergman NH, Phillippy AM. Interactive metagenomic visualization in a Web browser. BMC

Bioinformatics. 2011;**12**(1):289-294. Available from: http://www.biomedcentral.com/1471-2105/12/385

[30] Silva GGZ, Cuevas DA, Dutilh BE, Edwards RA. FOCUS: An alignment-free model to identify organisms in metagenomes using non-negative least squares. PeerJ. 2014;**2**:e425. DOI: 10.7717/peerj.425

[31] Wood DE, Salzberg SL. Kraken: Ultrafast metagenomic sequence classification using exact alignments. Genome Biology. 2014;**15**:R46. DOI: 10.1186/gb-2014-15-3-r46

[32] Langille MGI, Zaneveld J, Caporaso JG, McDonald D, Knights D, Reyes JA, et al. Predictive functional profiling of microbial communities using 16S rRNA marker gene sequences. Nature Biotechnology. 2013;**31**:814-821. DOI: 10.1038/nbt.2676

[33] Na SI, Kim YO, Yoon SH, min HS, Baek I, Chun J. UBCG: Up-to-date bacterial core gene set and pipeline for phylogenomic tree reconstruction. Journal of Microbiology. 2018;**56**:281-285. DOI: 10.1007/s12275-018-8014-6

[34] Lee MD. GToTree: A user-friendly workflow for phylogenomics. Bioinformatics. 2019. DOI: 10.1093/bioinformatics/btz188

[35] Sharpton TJ, Riesenfeld SJ, Kembel SW, Ladau J, O'Dwyer JP, Green JL, et al. PhyLOTU: A high-throughput procedure quantifies microbial community diversity and resolves novel taxa from metagenomic data. PLoS Computational Biology. 2011;**7**:e1001061. DOI: 10.1371/journal.pcbi.1001061

[36] Darling AE, Jospin G, Lowe E, Matsen FA, Bik HM, Eisen JA. PhyloSift: Phylogenetic analysis of genomes and metagenomes. PeerJ. 2014;**2**:e243. DOI: 10.7717/peerj.243

[37] Mori H, Maruyama T, Yano M, Yamada T, Kurokawa K. VITCOMIC2: Visualization tool for the phylogenetic composition of microbial communities based on 16S rRNA gene amplicons and metagenomic shotgun sequencing. BMC Systems Biology. 2018;**12**:S2. DOI: 10.1186/s12918-018-0545-2

[38] Seemann T. barrnap 0.8: Rapid ribosomal RNA prediction [Internet]. 2013. Available from: https://github.com/tseemann/barrnap [Accessed: 08 October 2019]

[39] Gouy M, Guindon S, Gascuel O. SeaView version 4: A multiplatform graphical user interface for sequence alignment and phylogenetic tree building. Molecular Biology and Evolution. 2010;**27**(2):221-224. Available from: http://www.ncbi.nlm.nih.gov/pubmed/19854763

Section 3

Applications of Metagenomics

Chapter 6

Microbial Community Structure and Metabolic Networks in Polar Glaciers

Eva Garcia-Lopez, Ana Maria Moreno and Cristina Cid

Abstract

Polar glaciers are inhabited by numerous microorganisms including representatives of bacteria, archaea, microeukaryotes, and viruses. Low temperature is a main factor when considering polar glaciers as extreme environments. However, desiccation, low nutrients availability, ultraviolet irradiation, and photoreactive chemistry do also significantly influence their challenging life. Glaciers are highly selective and confined habitats, which make them favorable environments for adaptation and speciation. Depending on the glacier area studied, microorganisms establish a vertical food chain, from the surface photosynthesizers in upper illuminated layers to chemoautotrophs and heterotrophs confined to the inner part. These regions are rich not only in biodiversity but also in new mechanisms of adaptation to the environment, since selection acts with a particular intensity. Glaciers are retreating in many areas of the world due to global warming. When glaciers have ultimately withdrawn, microorganisms play a main role, carrying out key processes in the development of soil and facilitating plant colonization. These features make them unique and interesting for the study and protection of the biological heritage. Metagenomics have allowed a deeper understanding of microbial ecology and function of polar glacier microbial communities. In this review, we present a complete analysis of the microbial diversity in these ecosystems and include a thorough overview of the metabolic potentials and biogeochemical cycles in polar glacier habitats.

Keywords: polar glacier, metagenomics, diversity, community structure, metabolic networks

1. Introduction

Polar glaciers have aroused great interest over the last year, and their study has increased since they are sentinels of climate change. Although both poles are extreme environments (in terms of low temperatures, high UV radiation, lack of light in winter and permanent solar radiation in summer, scarce nutrients, etc.), Arctic and Antarctic glaciers are very different. The North Pole is an ocean surrounded by land, while the South Pole is a continent surrounded by water. This distinction confers them very unique geographical and environmental characteristics.

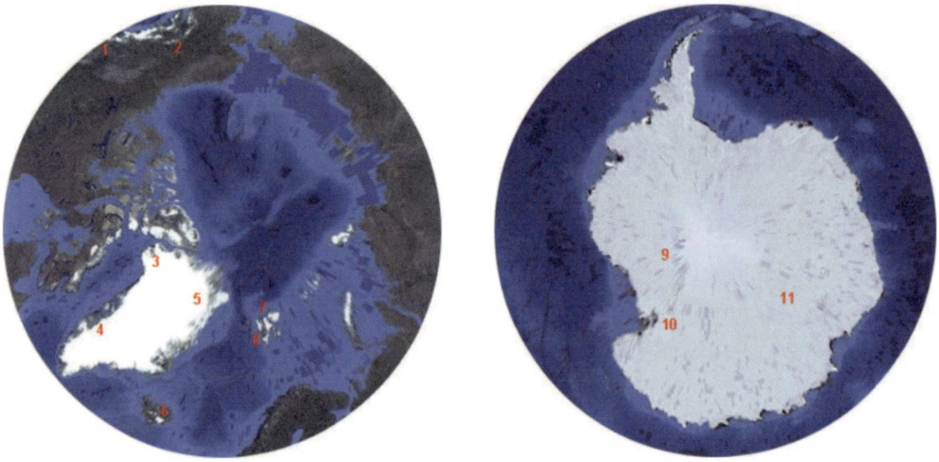

Figure 1.
Polar maps and localization of the referenced glaciers. (1) Cascade Volcano Arc [4]; (2) Robertson Glacier, Alberta [5]; (3) Western margin of the Greenland ice sheet [6]; (4) southwest part of the Greenland ice sheet [7]; (5) North-Eastern Greenland [8]; (6) Islandic glaciers [11, 12]; (7) Hamiltonbukta, Svalbard [9]; (8) Ny Ålesund, Svalbard [10]; (9) West Antarctic ice sheet [15]; (10) high Antarctic Plateau [13]; (11) Lake Vostok [14] (map source: Google Earth Pro).

Metabolically active microbial communities have been identified in both the Arctic [1] and Antarctic glaciers (**Figure 1**) [2]. These microbial communities include bacteria, archaea, microeukaryotes, and viruses [3].

Arctic glaciers do not reach such distant latitudes or low temperatures as Antarctic glaciers do. These are some of the reasons why they are being extremely affected by global warming. There are glaciers around the entire Arctic Ocean, but polar glaciers in North America [4, 5], Greenland [6–8], Svalbard [9, 10], and Iceland [11, 12] (**Figure 1**) have been the most widely studied from a microbiological point of view. Antarctic glaciers present exceptional environmental conditions. Being in higher latitudes allows the existence of very low temperatures and high rates of solar radiation in summer. Some published reports on glacial and subglacial microbiology refer to very extreme latitudes that reach −75°S in the high Antarctic Plateau [13], −77°S in Lake Vostok [14], and −84°S in the West Antarctic ice sheet [15] (**Figure 1**).

In the study of glacier microbiology, a variety of techniques have been traditionally used, such as microscopy techniques [16], cell cultures, and isolation of microorganisms [17]. However, the most significant advance has been achieved with the application of metagenomics. This discipline has allowed both the knowledge of the microbial communities' structure and the comprehension of their metabolic potential.

2. Microbial community structure through reconstruction of microbial genomes in polar glaciers

Glaciers have recently been considered authentic biomes [18]. It has been observed that microbial community composition depends on the area of the glacier studied [19]. In most of them, three well-defined and interconnected ecosystems can be defined: supraglacial, englacial, and subglacial ecosystems. These ecosystems are different in their solar radiation, water content, nutrient abundance, and redox potential [20]. These factors influence in the abundance and diversity of microbial populations inhabiting glaciers (**Figure 2**). They also affect the type of functionality and the biogeochemical cycles in these ecosystems.

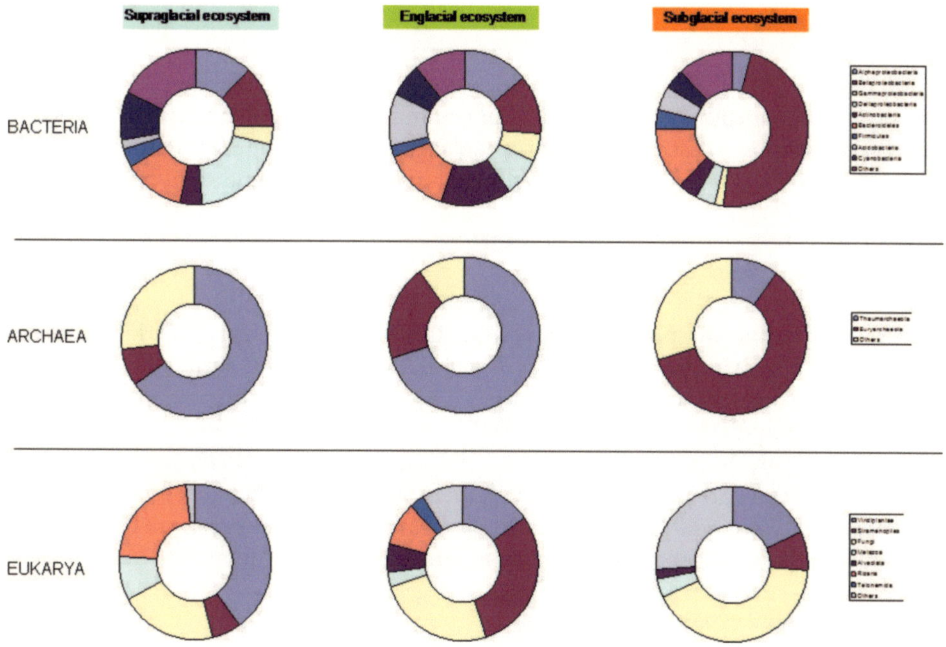

Figure 2.
Bacterial community structure in polar glacier ecosystems based on 16S and 18S rRNA gene sequences. Pie charts represent relative abundances of bacteria, archaea, and eukarya for three glacier ecosystems: supraglacial, englacial, and subglacial. The data are from [20, 11, 10, 4, 13, 8] for bacteria, from [11, 15, 5, 6, 14] for archaea, and from [19, 7, 12, 14] for eukarya.

2.1 The supraglacial ecosystem

The supraglacial ecosystem is the one which has best been studied. It has been reported that the main habitats in the supraglacial ecosystem are the snowpack, cryoconite holes (vertical cylindrical melt holes in a glacier surface), supraglacial streams, and moraines [20].

2.1.1 The glacial snow

The sunlit and oxygenated supraglacial surface is populated by autotrophic microorganisms such as cyanobacteria, microalgae, and diatoms [21], by chemolithotrophic bacteria, which feed on inorganic sand particles, and by heterotrophic bacteria and microeukaryotes [22] (**Figure 2**). The main bacterial classes that have been described in this ecosystem are *Betaproteobacteria*, *Actinobacteria*, and *Bacteroidetes* [11, 23]. The genus *Polaromonas* is one of the most abundant in the supraglacial area of Arctic [9] and Antarctic glaciers [24].

Among microeukaryotes, snow is mainly populated by pigmented algae, which have been observed in Arctic and Antarctic glaciers [25]. They belong to several taxa, mainly *Chlamydomonas*, *Chloromonas*, *Raphidonema*, and *Chrysophyceae* [11].

Fungi, especially basidiomycetous yeasts and *Chytridiomycota*, have been reported in glacial snow and ice [26]. It is believed that they act as saprophytes and parasites, yet their diversity and function in this ecosystem are poorly known [25].

Archaea have also been identified in glacial snow and ice, although they have not been found in all the studies that have been carried out. They belong to Nitrososphaerales, which are known as important ammonia oxidizers [11].

2.1.2 Ice surfaces

Under the recently fallen snow, there is a layer of hard ice. This layer arises to the surface in the episodes of melting that occur during the polar summer. The ice surface constitutes a distinct type of supraglacial microhabitat that is different from cryoconite holes. It is mainly populated by microalgae (Zygnematophyceae) and by cyanobacteria [17].

2.1.3 Cryoconite holes

Cryoconite holes are predominantly inhabited by cyanobacteria [27]. Filamentous cyanobacteria such as *Phormidesmis*, *Oscillatoria*, *Leptolyngbya*, *Phormidium*, and *Nostoc* play an important role in cryoconites [28]. They produce organic material and extracellular polymeric substances (EPS), which act as cryo- and osmo-protectants [27]. Additionally, bacteria of the class *Actinobacteria* (*Microbacteriaceae* and *Intrasporangiaceae*) are also important members of cryoconite holes, followed by *Proteobacteria*, *Bacteroidetes*, and *Cyanobacteria*. Archaea and eukarya are the least abundant and the least representative members of this environment [10].

2.2 The englacial ecosystem

In englacial ecosystems, live motile bacteria can reach more than 3000 m of depth. These bacteria reside in clay particles and ice channels. According to their metabolism, they can be both chemoautotrophs (i.e., *Streptomyces*, *Nocardia*, *Bacillus*) and heterotrophs (i.e., *Proteobacteria*, *Actinobacteria*) (**Figure 2**). The later bacteria feed on solubilized organic products from pollen grains and from other dead microorganisms. At great depth, anaerobic respiration takes place [29] and methanogens (for instance, *Firmicutes* and *Euryarchaeota*) are also active [20].

2.3 The subglacial ecosystem

The subglacial ecosystem is dominated by aerobic and anaerobic bacteria in basal bedrock and subglacial lakes. It does also contain diverse and metabolically active archaeal, bacterial, and fungal species [25] (**Figure 2**).

Among bacteria, species with chemolithotrophic activity have been identified; an example is *Sideroxydans lithotrophicus*, which is an iron sulfide oxidizer. Other bacterial taxa found in this ecosystem are *Thiobacillus* and *Thiomicrospira*, both associated with the sulfur and iron cycles [15].

Archaea in these anoxic environments are mainly represented by methanogenic and methanotrophic species [25]. Methanogenesis, the production of methane in an anaerobic process mediated exclusively by methanogenic archaea, is a very plausible process in the subglacial ecosystem. In glacier samples from this environment, methanogenic archaea of the euryarchaeal orders Methanosarcinales [5] and Methanomicrobiales have been detected [6].

Eukaryotes have only been found in some of the studied subglacial environments [19]. Among them, mainly fungi have been described [26]. *Basidiomycetes* predominate, among which *Cryptococcus* and *Rhodotorula* are the dominant genera.

3. Metabolic potentials and biogeochemical cycles in polar glaciers through reconstruction of microbial metagenomes

Living in such extreme environments implies coping with low temperatures, desiccation, low nutrients availability, and ultraviolet irradiation [30]. Over the last years, metagenomics have allowed a great understanding of metabolic potentials and biogeochemical cycles in polar glaciers through reconstruction of microbial genomes (**Figure 3**).

Regarding the supraglacial ecosystem, metagenomic studies have demonstrated the wide diversity of functions in cryoconite holes, with a range of metabolic pathways which depend on their competence to acquire and degrade available nutrients [10]. Functional analyses highlighted the importance of stress responses and efficient carbon and nutrient recycling.

Metagenomic techniques have also been used to identify algal communities in the supraglacial ecosystem and their relationship with geochemical factors [12].

The potential of archaea as important ammonia oxidizers has been another finding achieved by metagenomics [11].

Little is known about the metabolic potential and the biogeochemical cycles of microbial communities inhabiting the englacial ecosystem. It has been reported that microorganisms enclosed in the englacial ice present very low metabolic rates, using energy only to repair damaged biomolecules and not to grow and reproduce [31].

In the subglacial ecosystem, some metagenomics data implied that the most abundant and active component were bacteria within the order *Methylococcales* [6]. Transcripts of the particulate methane monooxygenase from these taxa were detected, demonstrating that methanotrophic bacteria were functional members of this subglacial ecosystem.

At least three modes of carbon fixation were inferred [14]. The most common mode of carbon fixation was the reductive pentose phosphate cycle. The second in frequency was the reductive tricarboxylic acid pathway. This cycle also produces

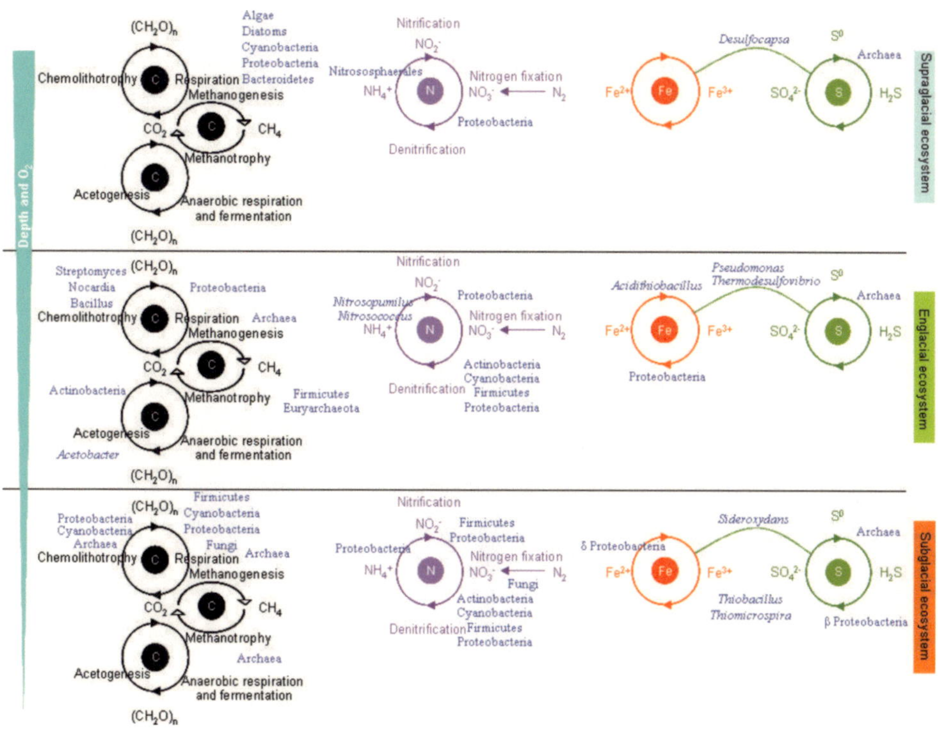

Figure 3.
Overview of the metabolic potentials between dominant microorganisms in the three polar glacial ecosystems. The data are from [20, 11, 10, 4, 13, 8] for bacteria, from [11, 15, 5, 6, 14] for archaea, and from [19, 7, 12, 14] for eukarya.

precursors for nucleic acid and aromatic amino acid syntheses. The third type of carbon fixation, the reductive acetyl-CoA pathway, is the one used by archaea [14].

These investigations did also identify genes that carry out various parts of the nitrogen cycle, including nitrogen fixation (*Actinobacteria*, *Cyanobacteria*, *Betaproteobacteria*, and *Gammaproteobacteria*), nitrification (*Alphaproteobacteria* and *Betaproteobacteria*), denitrification (*Gammaproteobacteria*), nitrate reduction (*Betaproteobacteria* and *Gammaproteobacteria*), anammox (*Planctomycetes*), assimilation (most microorganisms from these investigations), and decomposition (fungi and other heterotrophs) [14].

Characterization of the Antarctic Blood Falls microbial assemblage revealed taxa that could participate in active sulfur cycling, including autotrophs and heterotrophs such as *Desulfocapsa*, *Geopsychrobacter*, *Thiomicrospira*, and *Thermacetogenium* [32]. Although these microorganisms usually inhabit the subglacial ecosystem, in Blood Falls, they have been identified in brines collected from outflowing fluids (**Figure 3**).

4. Comparison of metagenome analysis techniques

The metagenome of polar microorganisms has been widely studied in recent years. Their results can provide a great amount of information about the biodiversity, survival capacity, and functioning of microbial communities in these extreme environments. In addition, information about ancient communities preserved within glacial ice through time can be obtained [33].

4.1 Sanger

Between 1975 and 2005, most of the DNA sequences were obtained through the application of the Sanger techniques [34], which led to the first generation of automated DNA sequencers [35] (**Figure 4**). For 16S or 18S rRNA sequencing, PCR amplification was carried out with specific primers (**Table 1**) and sequencing instruments based on capillary electrophoresis. Nowadays, Sanger sequencing achieves high read lengths of up to 1000 bp and per base and accuracies of 99.999%

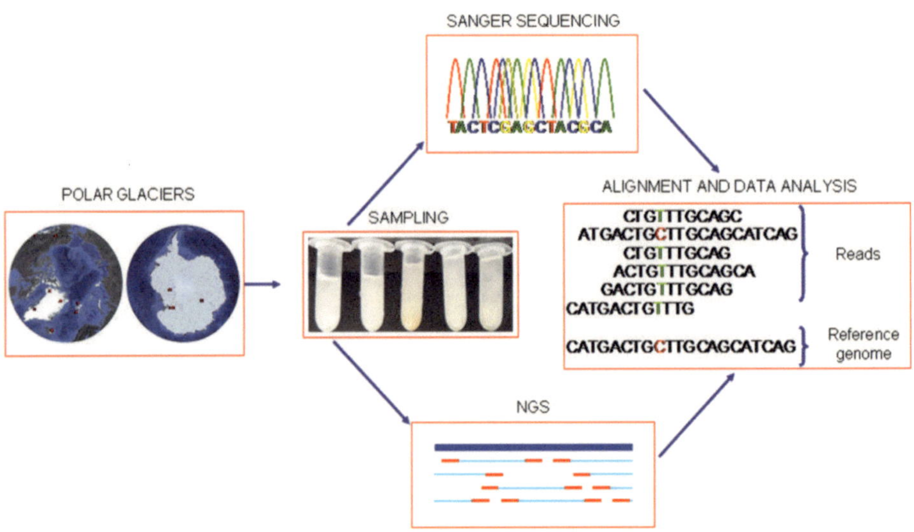

Figure 4.
Typical workflow in polar glacier metagenomics researches.

	Specificity	Primers	Sequence (5′ to 3′)	Product length (bp)	Authors	Reference
Sanger	Bacteria	16S-F	AGAGTTTGATCCTGGCTCAG	1000	Lane, 1991	[45]
	Bacteria	16S-R	CACGAGCTGACGACAGCC	1000	Lane, 1991	[45]
	Archaea	20F	TTCCGGTTGATCCYGCCRG	1372	Massana et al., 1997	[46]
	Archaea	U1392R	ACGGGCGGTGTGTRC	1372	Massana et al., 1997	[46]
	Eukarya	Euka1F	CTGGTTGATCCTGCCAG	500	Lefranc et al., 2005	[47]
	Eukarya	Euk502R	TGATCCTTCTGCAGGTTCACCTAC	500	Amann et al., 1990	[48]
NGS	Bacteria	(V3–V4) 341F	**ACACTGACGACATGGTTCTACA**CCTACGGGNGGCWGCAG	100	Herlemann et al., 2011	[49]
	Bacteria	(V3–V4) 805R	**TACGGTAGCAGAGACTTGGTCT**GACTACHVGGGTATCTAATCC	100	Herlemann et al., 2011	[49]
	Eukarya	(V9) 1380F	**GCCTCCCTCGCGCCATCAG**XXXXXCCCTGCCHTTTGTACACAC	43	Amaral-Zettler et al., 2009	[50]
	Eukarya	(V9) 1510R	**GCCTTGCCAGCCCGCTCAG**CCTTCYGCAGGTTCACCTAC	39	Amaral-Zettler et al., 2009	[50]

Table 1.
Primer sequences for 16S or 18S rRNA sequencing.

[36]. In the de novo metagenomics, randomly fragmented DNA was cloned into a high-copy-number plasmid and then transformed in *Escherichia coli*. However, whole-genome sequencing by this technology was extremely expensive and time consuming.

Some examples of microorganisms from polar glaciers analyzed with these technologies were Antarctic bacteria from the Dry Valleys [37] and the Arctic ice pack [38]. In general, the number of sequences identified by this technique was scarce. However, this method has the advantage of generating long reference sequences, which are very useful for studies of taxonomy and biodiversity.

4.2 NGS

Next-generation sequencing (NGS) technology is similar to capillary sequencing (**Figure 4**). The main difference is that, instead of sequencing a single DNA fragment, NGS develops this process with millions of DNA fragments.

The introduction of pyrosequencing technology by 454 life sciences in 2005 began the NGS innovation. This allowed the identification of thousands of short-sequencing reads without the need for cloning. This technique was used to research the microbial life in the Dry Valleys, Antarctica [39], and Ace Lake, Antarctica [40], and in Arctic glaciers from Svalbard [10].

Since then, many other NGS technologies have been developed. The Illumina platform (MiniSeq, MiSeq, NextSeq, HiSeq, and NovaSeq instruments) is based on sequencing by synthesis of the complementary strand and fluorescence-based detection of reversibly blocked terminator nucleotides. The platform includes multiple instruments with varying read length. For example, Illumina sequencing has been employed in a metagenomic research into diazotrophic communities across Arctic glacier forefields [41] and in the metagenomic analysis of basal ice from an Alaskan glacier [42]. Sequencing of 16S and 18S rDNA PCR amplicons is the most common approach to investigating environmental prokaryotic diversity, despite the known biases introduced during PCR. Recently this method has been improved with the use of 16S rDNA fragments derived from Illumina-sequenced environmental metagenomes [43]. Furthermore, newer Illumina sequencers produce longer reads (e.g., the HiSeq2500 and MiSeq produce 2 × 150bp and 2 × 250bp reads, respectively, which after merging can generate reads up to, e.g., 290 and 490 bp).

Other metagenomic studies based on the Ion Torrent platform were also based on sequencing by synthesis, but the detection was performed using semiconductor technology. Ion Torrent technology was applied to analyze red snow microbiomes and their role in melting Arctic glaciers [12].

The main drawback of the aforementioned second-generation sequencing platforms is that they generate relatively fragmented genome assemblies. In order to produce closed reference genomes, longer reads are required [36]. To meet this demand, third-generation sequencing platforms have been developed. These technologies directly target single DNA molecules without the need for PCR amplification. The PacBio RSII platform uses single-molecule real-time (SMRT) sequencing technology which allows to obtain extremely long DNA fragments of 20 kb and even longer [43].

4.3 Genome analysis tools

Environmental microbiome sequencing analysis consists of binning sequencing reads into taxonomic units to compare the microbial composition of samples. This information will allow the knowledge of the microbial population taxonomy, diversity, and functioning. When these data are correlated to certain environmental parameters,

Category	Tool[1]	Taxonomy[2]	Function	URL
Use all available sequences	MEGAN/MEGAN4	B	+	https://ab.inf.uni-tuebingen.de/software/megan4
	MG-RAST	B/E	+	https://www.mg-rast.org/
	Genometa	B	–	http://genomics1.mh-hannover.de/genometa/
	Kraken	B	–	https://ccb.jhu.edu/software/kraken/
	LMAT	B	+	https://computation.llnl.gov/projects/livermore-metagenomics-analysis-toolkit
	Taxator-tk	B	–	https://github.com/fungs/taxator-tk
	CLARK	B	–	http://clark.cs.ucr.edu/
	GOTTCHA	B/E	–	http://lanl-bioinformatics.github.io/GOTTCHA/
	EBI	B	+	https://www.ebi.ac.uk/metagenomics/
Use a set of genes	MetaPhyler	B	–	http://metaphyler.cbcb.umd.edu/
	QIIME6	B	–	http://qiime.org/
	mOTU	B	–	https://omictools.com/motu-tool
	MetaPhlAn	B/E	–	http://huttenhower.sph.harvard.edu/metaphlan2
	One Codex	B/E	–	https://onecodex.com/

[1]*Incomplete list compiled from sources.*
[2]*B, bacterial taxa; E, eukaryotic taxa.*

Table 2.
Metagenome analysis tools.

both ecological and biogeochemical analysis can be performed. Taxonomic binning of 16S and 18S rRNA reads is usually based on one of these four databases: SILVA, Ribosomal Database Project, Greengenes, and NCBI [44]. For instance, the Ribosomal Database Project was used to perform a metagenomic analysis if Illumina sequences to identify bacterial communities in Antarctic surface snow [45].

Several tools have been developed to investigate the taxonomic composition of metagenomes and, in some cases, the functional composition of the community. These tools can be classified into two groups: those that use all the available sequences (MEGAN/MEGAN4, MG-RAST, Genometa, Kraken, LMAT, Taxator-tk, CLARK, GOTTCHA, EBI) and those that use a set of genes (MetaPhyler, QIIME6, mOTU, MetaPhlAn, One Codex) [33]. These genome analysis tools are summarized in **Table 2**.

An example of the use of these tools is the metagenomics analysis with MG-RAST performed to study Arctic microbial communities [41]. Sequence analysis with QIIME was performed with cryoconite samples from Arctic glaciers [10] and with permafrost samples from the Antarctic Dry Valleys [39].

Although metagenomics is changing rapidly, still new improvements in the development of analytical tools and databases are required to answer important questions in polar glacier microbiology.

5. Conclusions

- Extraordinary advances in metagenomics have allowed a great understanding of microbial ecology and function of polar glacier microbial communities.

- Important novel tools to study environmental microbiology based on metagenomics are being developed.
- Third-generation technologies may further revolutionize metagenomic research.
- The application of new technologies to metagenomic studies of polar glaciers will enable to link the diversity and functionality of these habitats.
- Significant challenges for metagenomics remain, especially in data processing and genome analysis.

Acknowledgements

This research was supported by the Spanish Ministerio de Ciencia, Innovación y Universidades (Grants CTM 2010-12134-E/ANT and CTM2011-16003-E) and the Institute of Health Carlos III (Grant PI14/00705). E.G.L. is a recipient of a Ciencia, Innovación y Universidades Fellowship (Grant PTA2016-12325-I, Programa Nacional de Contratación e Incorporación de RRHH). We thank Paula Alcazar for reading this manuscript and for her helpful suggestions and revisions.

Conflict of interest

The authors declare that none of us has any competing commercial interests in relation to the submitted work.

Author details

Eva Garcia-Lopez, Ana Maria Moreno and Cristina Cid*
Microbial Evolution Laboratory, Center for Astrobiology (CSIC-INTA), Madrid, Spain

*Address all correspondence to: cidsc@inta.es

References

[1] Skidmore ML, Foght JM, Sharp MJ. Microbial life beneath a high Arctic glacier. Applied and Environmental Microbiology. 2000;**66**:3214-3220

[2] Foght J, Aislabie J, Turner S, Brown CE, Ryburn J, Saul DJ, et al. Culturable bacteria in subglacial sediments and ice from two southern hemisphere glaciers. Microbial Ecology. 2004;**47**:329-340. DOI: 10.1007/s00248-003-1036-5

[3] Garcia-Lopez E, Cid C. The role of microbial ecology in glacier retreat analysis. In: Tangborn WV, editor. Glaciers. Rijeka, Croatia: InTech; 2017. DOI: 10.5772/intechopen.69097

[4] Hamilton TL, Havig J. Primary productivity of snow algae communities on stratovolcanoes of the Pacific Northwest. Geobiology. 2017;**15**: 280-295. DOI: 10.1111/gbi.12219

[5] Boyd ES, Hamilton TL, Havig JR, Skidmore ML, Shock EL. Chemolithotrophic primary production in a subglacial ecosystem. Applied and Environmental Microbiology. 2014;**80**:6146-6153. DOI: 10.1128/AEM.01956-14

[6] Dieser M, Broemsen EL, Cameron KA, King GM, Achberger A, Choquette K, et al. Molecular and biogeochemical evidence for methane cycling beneath the Western margin of the Greenland ice sheet. The ISME Journal. 2014;**8**:2305-2316. DOI: 10.1038/ismej.2014.59

[7] Yallop ML, Anesio AM, Perkins RG, Cook J, Telling J, Fagan D, et al. Photophysiology and albedo-changing potential of the ice algal community on the surface of the Greenland ice sheet. The ISME Journal. 2012;**6**:2302-2313. DOI: 10.1038/ismej.2012.107

[8] Møller AK, Søborg DA, Al-Soud WA, Sørensen SJ, Kroer N. Bacterial community structure in high-arctic snow and freshwater as revealed by pyrosequencing of 16S rRNA genes and cultivation. Polar Research. 2013;**32**:17390. DOI: 10.3402/polar.v32i0.17390

[9] Garcia-Lopez E, Rodriguez-Lorente I, Alcazar P, Cid C. Microbial communities in coastal glaciers and tidewater tongues of Svalbard archipelago, Norway. Frontiers in Marine Science. 2019;**5**:512. DOI: 10.3389/fmars.2018.00512

[10] Edwards A, Mur LA, Girdwood SE, Anesio AM, Stibal M, Rassner SM, et al. Coupled cryoconite ecosystem structure-function relationships are revealed by comparing bacterial communities in alpine and Arctic glaciers. FEMS Microbiology Ecology. 2014;**89**:222-237. DOI: 10.1111/1574-6941.12283

[11] Lutz S, Anesio AM, Edwards A, Benning LG. Microbial diversity on Icelandic glaciers and ice caps. Frontiers in Microbiology. 2015;**6**:307. DOI: 10.3389/fmicb.2015.00307

[12] Lutz S, Anesio AM, Raiswell R, Edwards A, Newton RJ, Gill F, et al. The biogeography of red snow microbiomes and their role in melting arctic glaciers. Nature Communications. 2016;**7**:11968. DOI: 10.1038/ncomms11968

[13] Michaud L, Lo Giudice A, Mysara M, Monsieurs P, Raffa C, Leys N, et al. Snow surface microbiome on the high Antarctic plateau (DOME C). PLoS One. 2014;**9**:e104505. DOI: 10.1371/journal.pone.0104505

[14] Shtarkman YM, Kocer ZA, Edgar R, Veerapaneni RS, D'Elia T, Morris PF, et al. Subglacial Lake Vostok (Antarctica) accretion ice contains a diverse set of sequences from aquatic, marine and sediment-inhabiting bacteria and Eukarya. PLoS One. 2013;**8**:e67221. DOI: 10.1371/journal.pone.0067221

[15] Christner BC, Priscu JC, Achberger AM, Barbante C, Carter SP, Christianson K, et al. A microbial ecosystem beneath the West Antarctic ice sheet. Nature. 2014;**512**:310-313. DOI: 10.1038/nature13667

[16] Deming JW. Psychrophiles and polar regions. Current Opinion in Microbiology. 2002;**5**:301-309

[17] Martín-Cerezo ML, García-López E, Cid C. Isolation and identification of a red pigment from the Antarctic bacterium *Shewanella frigidimarina*. Protein and Peptide Letters. 2015;**22**:1076-1082

[18] Anesio AM, Laybourn-Parry J. Glaciers and ice sheets as a biome. Trends in Ecology & Evolution. 2012;**27**:219-225. DOI: 10.1016/j.tree.2011.09.012

[19] Boetius A, Anesio AM, Deming JW, Mikucki JA, Rapp JZ. Microbial ecology of the cryosphere: Sea ice and glacial habitats. Nature Reviews. Microbiology. 2015;**13**:677-690. DOI: 10.1038/nrmicro3522

[20] Hodson A, Anesio AM, Tranter M, Fountain A, Osborn M, Priscu J, et al. Glacial ecosystems. Ecological Monographs. 2008;**78**:41-67

[21] Garcia-Lopez E, Cid C. Glaciers and ice sheets as analog environments of potentially habitable icy worlds. Frontiers in Microbiology. 2017;**8**:1407. DOI: 10.3389/fmicb.2017.01407

[22] Mieczen T, Gorniak D, Swiatecki A, Zdanowski M, Tarkowska-Kukuryk M, Adamczuk M. Vertical microzonation of ciliates in cryoconite holes in ecology glacier, King George Island. Polish Polar Research. 2013;**34**:201-212. DOI: 10.2478/popore-2013-0008

[23] Xiang S, Shang T, Chen Y, Jing Z, Yao T. Dominant bacteria and biomass in the Kuytun 51 glacier. Applied and Environmental Microbiology. 2009;**75**:7287-7290. DOI: 10.1128/AEM.00915-09

[24] Irgens RL, Gosink JJ, Staley JT. *Polaromonas vacuolata* gen. nov., sp. nov., a psychrophilic, marine, gas vacuolate bacterium from Antarctica. International Journal of Systematic Bacteriology. 1996;**46**:822-826. DOI: 10.1099/00207713-46-3-822

[25] Anesio AM, Lutz S, Chrismas NAM, Benning LG. The microbiome of glaciers and ice sheets. NPJ Biofilms and Microbiomes. 2017;**3**:10. DOI: 10.1038/s41522-017-0019-0

[26] Butinar L, Spencer-Martins I, Gunde-Cimerman N. Yeasts in high Arctic glaciers: The discovery of a new habitat for eukaryotic microorganisms. Antonie van Leeuwenhoek. 2007;**91**:277-289. DOI: 10.1007/s10482-006-9117-3

[27] Stibal M, Sabacka M, Kastovska K. Microbial communities on glacier surfaces in Svalbard: Impact of physical and chemical properties on abundance and structure of cyanobacteria and algae. Microbial Ecology. 2006;**52**:644-654. DOI: 10.3389/fmicb.2015.00225

[28] Chrismas NA, Anesio AM, Sánchez-Baracaldo P. Multiple adaptations to polar and alpine environments within cyanobacteria: A phylogenomic and Bayesian approach. Frontiers in Microbiology. 2015;**6**:1070. DOI: 10.3389/fmicb.2015.01070

[29] Price PB. Microbial life in glacial ice and implications for a cold origin of life. FEMS Microbiology Ecology. 2007;**59**:217-231

[30] Garcia-Lopez E, Alcazar P, Postigo M, Cid C. The effect of climate change on microbial communities from glaciers. In: Glaciers: Formation, Climate Change and Their Effects. New York: Nova Science Publishers, Inc; 2016. pp. 71-88

[31] Tung HC, Bramall NE, Price BP. Microbial origin of excess methane in glacial ice and implications for life on Mars. PNAS. 2005;**102**:18292-18296. DOI: 10.1073/pnas.0507601102

[32] Mikucki JA, Pearson A, Johnston DT, Turchyn AV, Farquhar J, Schrag DP, et al. A contemporary microbially maintained subglacial ferrous "ocean". Science. 2009;**324**:397-400. DOI: 10.1126/science.1167350

[33] Lindgreen S, Adair KL, Gardner PP. An evaluation of the accuracy and speed of metagenome analysis tools. Scientific Reports. 2016;**6**:19233. DOI: 10.1038/srep19233

[34] Sanger F, Nicklen S, Coulson AR. DNA sequencing with chain-terminating inhibitors. PNAS. 1977;**74**(12):5463-5467

[35] Garcia-Descalzo L, Garcia-Lopez E, Postigo M, Baquero F, Alcazar A, Cid C. Eukariotic microorganisms in cold environments: Examples from Pyrenean glaciers. Frontiers in Microbiology. 2013;**4**:55. DOI: 10.3389/fmicb.2013.00055

[36] Shendure J, Ji H. Next-generation DNA sequencing. Nature Biotechnology. 2008;**26**:1135-1145. DOI: 10.1038/nbt1486

[37] Bidle KD, Lee S, Marchant DR, Falkowski PG. Fossil genes and microbes in the oldest ice on earth. PNAS. 2007;**104**:13455-13460. DOI: 10.1073/pnas.0702196104

[38] Brinkmeyer R, Glöckner FO, Helmke E, Amann R. Predominance of ß-proteobacteria in summer melt pools on Arctic pack ice. Limnology and Oceanography. 2004;**49**:1013-1021. DOI: 10.4319/lo.2004.49.4.1013

[39] Goordial J, Davila A, Lacelle D, Pollard W, Marinova MM, Greer CW, et al. Nearing the cold-arid limits of microbial life in permafrost of an upper dry valley, Antarctica. The ISME Journal. 2016;**10**:1613-1624. DOI: 10.1038/ismej.2015.239

[40] Lauro FM, DeMaere MZ, Yau S, Brown MV, Ng C, Wilkins D, et al. An integrative study of a meromictic lake ecosystem in Antarctica. The ISME Journal. 2011;**5**:879-895. DOI: 10.1038/ismej.2010.185

[41] Nash MV, Anesio AM, Barker G, Tranter M, Varliero G, Eloe-Fadrosh EA, et al. Metagenomic insights into diazotrophic communities across Arctic glacier forefields. FEMS Microbiology Ecology. 2018;**94**:9. DOI: 10.1093/femsec/fiy114

[42] Kayani MUR, Doyle SM, Sangwan N, Wang G, Gilbert JA, Christner BC, et al. Metagenomic analysis of basal ice from an Alaskan glacier. Microbiome. 2018;**6**:123. DOI: 10.1186/s40168-018-0505-5

[43] Besser J, Carleton HA, Gerner-Smidt P, Lindsey RL, Trees E. Next-generation sequencing technologies and their application to the study and control of bacterial infections. Clinical Microbiology and Infection. 2018;**24**:335-341. DOI: 10.1016/j.cmi.2017.10.013

[44] Balvočiūtė M, Huson DH. SILVA, RDP, Greengenes, NCBI and OTT—How do these taxonomies compare? BMC Genomics. 2017;**18**:114. DOI: 10.1186/s12864-017-3501-4

[45] Lane DJ. 16S/23S rRNA sequencing. In: Stackebrandt E, Goodfellow M, editors. Nucleic Acid Techniques in Bacterial Systematic. New York: John Wiley and Sons; 1991. pp. 115-175

[46] Massana R, Murray AE, Preston CM, DeLong EF. Vertical distribution and phylogenetic characterization of marine planktonic Archaea in the Santa Barbara channel. Applied

and Environmental Microbiology. 1997;**63**:50-56

[47] Lefranc M, Thenot A, Lepere C, Debroas D. Genetic diversity of small eukaryotes in lakes differing by their trophic status. Applied and Environmental Microbiology. 2005;**71**:5935-5942. DOI: 10.1128/AEM.71.10.5935-5942.2005

[48] Amann RI, Binder BJ, Olson RJ, Chisholm SW, Devereux R, Stahl DA. Combination of 16S rRNA-targeted oligonucleotide probes with flow cytometry for analyzing mixed microbial populations. Applied and Environmental Microbiology. 1990;**56**:1919-1925

[49] Herlemann DP, Labrenz M, Jurgens K, Bertilsson S, Waniek JJ, Andersson AF. Transitions in bacterial communities along the 2000 km salinity gradient of the Baltic Sea. The ISME Journal. 2011;**5**:1571-1579. DOI: 10.1038/ismej.2011.41

[50] Amaral-Zettler LA, Mc Climent EA, Ducklow HW, Huse SM. A method for studying protistan diversity using massively parallel sequencing of V9 hyper variable regions of small-subunit ribosomal RNA genes. PLoS One. 2009;**4**:e6372. DOI: 10.1371/journal.pone.0006372

Chapter 7

Dissemination of Intestinal Microbiota by Migratory Birds across Geographical Borders

Takehiko Kenzaka

Abstract

Understanding the dissemination of medically important microbiota is a significant public health necessity. Although modern bacteriology has advanced with improved culturing technology, several environmental bacteria occur in the "viable but nonculturable" state. By using next-generation sequencing (NGS) to comprehensively analyze the intestinal microbiota of migratory birds, research on microbial communities traveling over long distances has entered a new era and provides new insights that are valuable for the analysis of medical care, livestock industry, agriculture, and human health risks. The use of comprehensive analysis by NGS of not only intestinal microbiota but also diet biological communities may help elucidate the relationship between microbiological communities and the diet and succession of intestinal microbiota, including antibiotic-resistant bacteria, during migration and breeding. Here, we have described the current state and the future implications of studying intestinal microbiota associated with migratory birds.

Keywords: migratory bird, avian, gut microbiota, intestinal microbiota, antibiotic resistance, colistin

1. Introduction

The application of DNA sequence technology covers a wide range of fields. As next-generation sequencing (NGS) has progressed, it has become more widely used in an array of practical applications [1, 2]. One direction of use includes the field of precision medical care. In cases of cancer caused by genetic mutations, molecular targeted drugs can be discovered by investigating gene mutations. In addition, the application of NGS is advancing in industrial fields; for instance, genotyping of animals and plants can be performed at low cost and the genetic markers can be screened with NGS.

NGS enables profiling of complex microbial communities in nature as well as that of indigenous microbiota associated with living organisms [3, 4]. Higher forms of life coexist with huge numbers of microorganisms, including bacteria, viruses, fungi, and, in some cases, protozoa and parasites—albeit bacteria are the most important microorganisms in terms of their numbers and host interactions. There are as many viruses as numbers, but many of them are viruses (phages) that infect bacteria.

Bacteria are present on the surface bodies of living organisms, such as their skin, gastrointestinal tracts, respiratory systems, and oral cavities, and they are colonized with an inherent balance in each place. This balance constitutes a stable complex ecosystem through crosstalk between bacteria and hosts. Among these places of localization, the digestive tract has the most abundant localization in terms of both the number and the type. In humans, 90% of established bacteria inhabit the digestive tract.

The intestinal microbial community includes not only enormous numbers and types of microorganisms but also active metabolic activity. The genes of the intestinal microbial community are present in proportion of at least 100-fold more than that of the genome of the host and various metabolites are produced, which are absorbed into the host body. NGS is now an indispensable item in the study of intestinal microbial communities and is applied not only to humans but also to other living organisms, such as domestic animals, insects, poultry, and wild birds [5].

Understanding the transboundary movement of microorganisms is an important requirement from the perspective of public health and environmental science [6, 7]. Microorganisms travel geographically over distant areas on the earth via ocean currents and atmospheric movement. In addition, migratory birds carry pathogenic microorganisms when traveling over long distances to several parts of the world [8–10]. Numerous pathogenic bacteria, such as pathogenic *Escherichia coli*, *Salmonella* spp., *Listeria monocytogenes*, *Pseudomonas aeruginosa*, *Botulinum* spp., *Listeria* spp., and *Campylobacter jejuni*, have been isolated from bird feces [11–13].

Research efforts have been performed to identify the bacterial communities contained in the feces of migratory birds (Bar-headed goose, shorebirds, swallow, etc.) by a novel culture-independent method [14, 15]. We attempted to explore the stability of the intestinal bacterial communities in migratory birds, the difference in the intestinal bacterial communities among birds at the individual and species levels, and the potential of long-distance movement of antibiotic-resistant bacteria associated with migratory birds [16–19]. Research on the spread of bacterial populations over vast distances has led to the elucidation of the roles of migratory birds regarding human health risks, thereby enabling the prediction of potential outbreaks based on their migratory patterns. NGS is useful for understanding bacterial diversity and for discovering novel bacteria [20]. The present review considers the potential role of wild birds in the transmission of intestinal microbiota, including antibiotic-resistant microorganisms, and our current knowledge of microbiota associated with migratory birds using NGS technologies.

2. Methodology for analysis of bacterial community composition

Since the scientific study on bacteria began in the 19th century, pure culture methods supported the progress of microbiology in a wide range of fields such as medicine, pharmacy, biology, agriculture, and fermentation engineering. In the latter half of the 1970s, however, the method of total direct counting was developed, in which bacteria were stained with fluorescent dye and directly observed or counted under a fluorescence microscope. The use of this method revealed that several environmental bacteria cannot yet be cultivated by the conventional laboratory techniques [21, 22]. Therefore, new bacterial detection methods that were independent of culturing began to be developed in succession [23–26]. From the latter half of the 1990s, the method to directly extract DNA from the sample without culturing the bacteria and using a universal primer to target the conserved region of the 16S rRNA gene or PCR amplification with genus-specific primers to decode the DNA sequence became widespread [15, 27, 28]. Since the obtained gene information depended on the number of bacterial clones that the researcher could handle at a time, these methods were found to be limited to about tens to thousands.

In the past few years, comprehensive analysis of DNA sequences using NGS has spread rapidly [5]. NGS is a powerful fundamental technology that is capable of concurrently determining the nucleotide sequences of tens of millions to hundreds of millions of DNA fragments. It is also capable of advanced and high-speed processing, such as multiple determinations of multiple samples. Moreover, the expenditure on equipment and operations for this method has also been reduced. The use of NGS can help acquire genetic information of tens of thousands to hundreds of thousands of bacterial species in a short time. NGS can also aid in the understanding of the entire picture, thereby enabling a greater focus on specific interesting bacteria based on the phylogenetic taxonomic information. This, in turn, would lead to further qualitative and quantitative analyses in detail.

Because 16S rRNA gene contains both highly conserved regions for primer design and hypervariable regions to identify the phylogenetic characteristics of microorganisms, 16S rRNA gene sequence has become the most widely used marker gene for profiling bacterial communities [29, 30]. Full-length 16S rRNA gene sequences consist of nine hypervariable regions that are separated by nine highly conserved regions [31, 32]. Study with bioinformatics tools attempted to evaluate the phylogenetic sensitivity of the hypervariable regions in comparison with the corresponding full-length sequences and revealed that the V4–V6 regions represented the optimal subregions for bacterial phylogenetic studies of the new phyla [33]. Since the 16S rRNA gene differs from 1 to 16 in the number of copies per cell depending on the genus [34], the relative proportion obtained by NGS does not necessarily agree with the ratio of actual community composition, although the dominant populations can be ascertained.

3. Migratory birds and flyway

The geographical route that migratory birds move annually on the earth is called "flyway," and there are nine major flyways in the world. Japan is located on the East Asia-Australia flyway, and it is estimated that more than 50 million migratory birds, such as shorebirds, birds, and seabirds, travel over the flyway every year.

Summer migrants in Japan fly from the south mainly for breeding and spend the summer in Japan, and when the breeding season ends, they return south for overwintering. Winter migrants in Japan fly from the north mainly for overwintering and spend the winter in Japan, and then in the spring, they return north for breeding. Passage migrants breed in the country north of Japan and overwinter in the country south of Japan, and so they travel through Japan during the movement and are mainly observed in spring and autumn. As a whole, some millions of migratory birds are estimated to visit Japan annually.

The direct counting method by fluorescent staining revealed that the fecal matter of migratory birds contains $\geq 10^8$ cells/g bacteria [16, 17]. Since disinfection treatments for bird feces is not performed like that for humans and livestock, there are possibilities that several live bacteria, along with 50 million migratory birds each year, travel the East Asia–Australia region flyway for a long distance. In order to clarify the dynamic of the microbiota, including pathogenic bacteria and antibiotic-resistant bacteria, as well as to verify their significance in public health and environmental microbiology, research has been performed to analyze the intestinal microbial community associated with migratory birds by NGS [16–22].

4. Intestinal microbiota of adult and young barn swallows

The barn swallow (*Hirundo rustica*) is about 17 cm in length and is widely distributed across the temperate and cold regions of Africa, the Eurasian continent, the northern end of Australia, and North America. Their global population is estimated at more than 190 million individuals [35]. Several swallows migrate to Japan from Southeast Asia (i.e., the Philippines, Malaysia, and Indonesia) and make their nests by mixing mud with plant pieces, feathers, and other such items near human living environments, such as the eaves of a house. While flying in the sky, the barn swallows catch insects to eat. After breeding, they return to Southeast Asia in the autumn season. The number of observed individuals in Japan is estimated at several hundred thousand birds per year.

For the study on the intestinal microbial communities of barn swallow, fecal specimens of the adult and young barn swallows were collected from the nest on the university campus and from various nests around the Osaka prefecture, Japan. The NGS analysis was performed on the V4 region of the 16S rRNA gene.

Diversity indices of intestinal bacterial communities in barn swallows were determined for each of the sampling locations (**Table 1**). The diversity index of young birds was found to be lower than that of adult birds. Similar results were reported from studies with swallows in Europe [36].

The stability of intestinal bacterial communities in wild birds has not yet been studied. We collected feces over a period of time from the same nest within the university campus (OOU 20) and compared changes in the intestinal bacterial communities between adult and young birds. The day when the birds were born was set as day 0, that before the birth as day –N, and that after the birth as day +N (where, N = 1, 2, 3, ...; **Figure 1**). Fecal samples in negative and positive value days represent ones of parent and young birds, respectively.

Adult samples were dominated by *Corynebacteriaceae*, *Halomonadaceae*, and *Pseudomonadaceae* at the family level before the hatch. In contrast, young samples were dominated by *Enterococcaceae*, *Mycoplasmataceae*, and *Enterobacteriaceae* (**Figure 1a**).

Figure 1b depicts the similarity in the intestinal bacterial community at the family level by principal component analysis (PCA). The intestinal bacterial communities of the adult birds changed greatly over time, but the profiles of adult and young birds remained similar before and after hatching, and then the young birds showed a great change. The adult birds brought insects and the likes in the nest to feed the young ones; therefore, it seems that the intestinal bacterial community was temporarily similar because of the similar diets. Similar results were reported from studies with swallows in Europe [36].

To examine the extent of change in the intestinal bacterial community of the barn swallows living in the same nest, the change was compared with that in other nests. The sampling sites OOU17 and OOU20 are nests located about 50 m apart.

Site	Adult	Young
All	2.48 (1.09)[a]	1.66 (0.85)
OOU17	2.74 (0.92)	2.34 (0.88)
OOU20	3.33 (1.36)	1.82 (0.76)
Wakayama	1.86 (0.40)	1.40 (0.73)

[a]*Standard deviation.*

Table 1.
Alpha diversity of fecal microbiota in adult and young barn swallow.

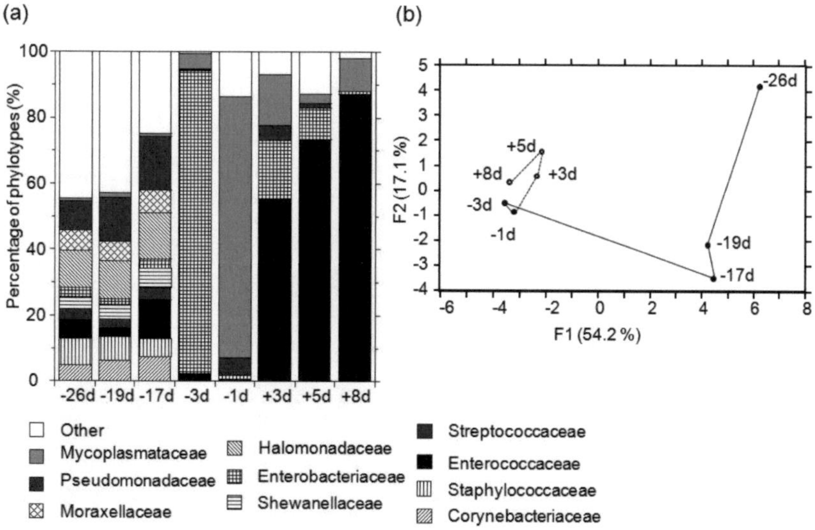

Figure 1.
Change in the bacterial community composition of fecal samples of barn swallows living in the same nest. (a) Relative portions of bacterial phylotypes in averaged fecal samples at the family level (b) Principle component analysis of family abundance data. Fecal samples in negative and positive value days represent ones of parent and young birds, respectively.

The sampling sites Wakayama were about 50 km away from the sites OOU20 and OOU17 and consisted of a plurality of nests at a distance of 100 m. As compared with these adult and young birds, no such characteristics were observed in the groups of adult or young birds or in specific nests (**Figure 2**). As seen in the figure regarding the change in the bacterial community at OOU 20, the extent of difference in the bacterial community over a period of time in the same nest was found to exceed the extent of the difference among different nests.

In order to verify whether there were similarities in the bacterial communities for each collection area, PCA was performed on those from the adult birds (**Figure 3a** and **b**). We examined four areas (northern Osaka, southern Osaka, northern Wakayama, and others); these areas were about 50 km apart and beyond the activity range of the insects that served as the swallows' diets. The distribution of the samples in the same nest at OOU20 is shown in gray in **Figure 3a**. The extent of difference in the bacterial community in the same nest surpassed that

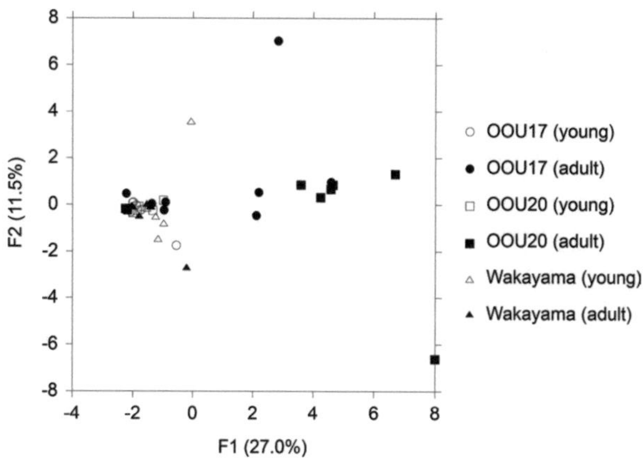

Figure 2.
Principal component analysis of class abundance data from adult and young barn swallows.

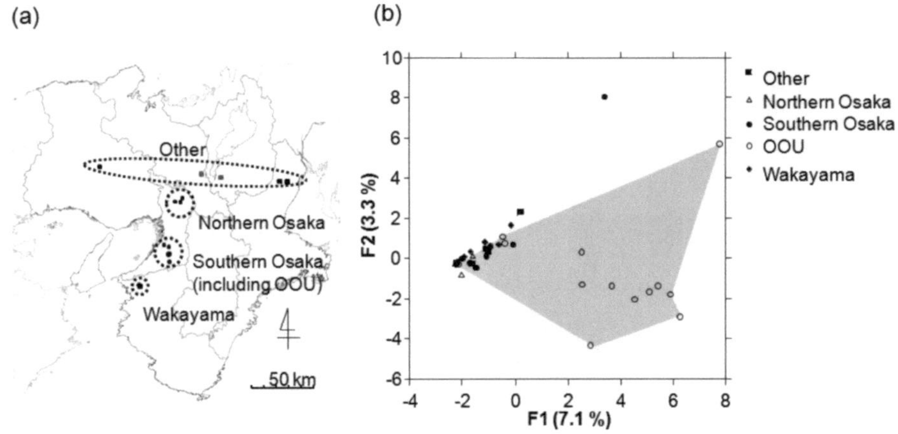

Figure 3.
Similarity analysis of 16S rRNA gene data from adult barn swallows. (a) Sampling sites around Osaka, Japan. (b) Principal component analysis of class abundance data.

among different nests. No relationship was found between the bacterial community composition and the geographical area in the fecal samples. Since the intestinal tract of swallows is short, it was speculated that the intestinal microbial community might be influenced by the daily diet and may accordingly change greatly in only a couple of days.

5. Succession of intestinal microbiota of a Eurasian wigeon while spending a winter

In order to examine the stability of the intestinal bacterial communities of other birds, we examined the succession of intestinal bacterial communities in the feces of a Eurasian wigeon, which was flying to the northern part of Osaka at the beginning of the winter season, staying in the same area when spending a winter. The Eurasian wigeon (*Mareca penelope* or *Anas penelope*) is about 50 cm in length. It breeds in the northern part of the Eurasian Continent, and in winter, it crosses southern Europe, North Africa, and East/South Asia. The global population of Eurasian wigeons is estimated at 2.8–3.3 million individuals [37]. This bird lives primarily in quiet sea, estuaries, lakes, and rivers. In addition to eating plants such as grass leaves and algae, it eats aquatic insects and mollusks. The number of wigeons observed in Japan is reportedly 180,000 per year.

In December 2017, when the researchers flew to Japan, we examined the succession over time with monthly intestinal bacteria of Eurasian wigeon around the Ai River in north Osaka (**Figure 4**). Community analysis at the class level revealed that Clostridia constituted 64.7% in December, but the proportion decreased to 18.4% in April 2018 ($P < 0.01$). The proportion of Bacilli, Fusobacteria, Alphaproteobacteria, and Gammaproteobacteria significantly increased from 0.3 to 7.4%, 1.2 to 10.4%, 1.2 to 6.3%, and 4.1–29%, respectively ($P < 0.01$), while spending a winter in Japan. The intestinal bacterial community composition of Eurasian wigeon, which flew to Japan at the beginning of the winter season, significantly changed while staying in Japan for about 4 months.

We also compared the intestinal microbial communities of Eurasian wigeon in December and April at different sites and in different years. We used samples collected from different sampling sites, Lake Biwa in 2016 and Ai River in 2017, as samples in the early winter. As samples for the spring season, fecal samples were collected from Biwa Lake and Ai River between 2016 and 2018.

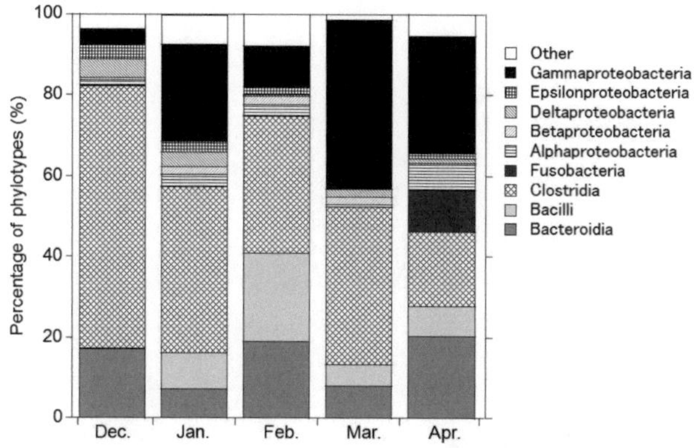

Figure 4.
Changes in the bacterial community composition of fecal samples from the Eurasian wigeon while staying in western Japan. Fecal samples were collected from around Ai River, north Osaka, Japan.

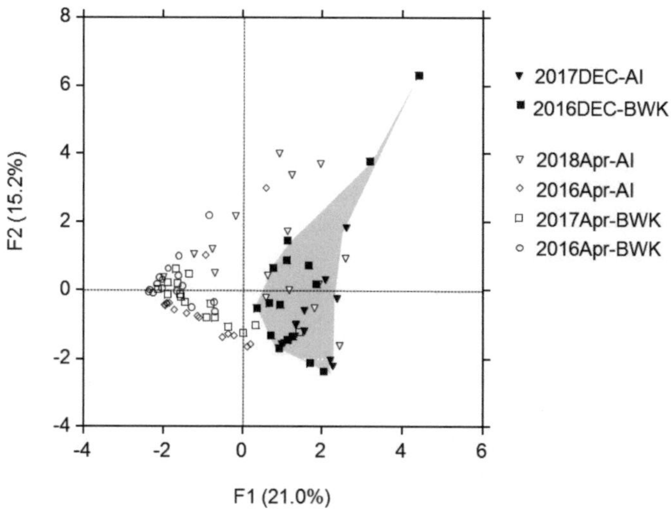

Figure 5.
Comparison of the bacterial community composition of fecal samples from the Eurasian wigeon in the early winter (December) and spring (April) seasons. Fecal samples were collected from around Ai River and lake Biwa, Japan in different years.

All samples in December were dominated by Clostridia, and samples in April were dominated by Gammaproteobacteria. Regardless of the differences in the location and year of sampling, the major bacterial communities were found to be similar. PCA revealed obvious difference in the bacterial community—namely, the samples in December were distributed in gray (right panel) and the samples in April shifted to the left panel in **Figure 5**. These results revealed that the intestinal bacterial community composition of Eurasian wigeon, which flew to Japan at the beginning of winter, changed while they spend a winter in Japan.

6. Colistin-resistant bacteria associated with Eurasian wigeon

The dissemination of antibiotic-resistant bacteria across borders has become an important issue in public health, and the involvement of migratory birds has been indicated as a mechanism by which resistant bacteria spread at the global level [13]. In addition, colistin is regarded as an important antimicrobial agent; it is even an

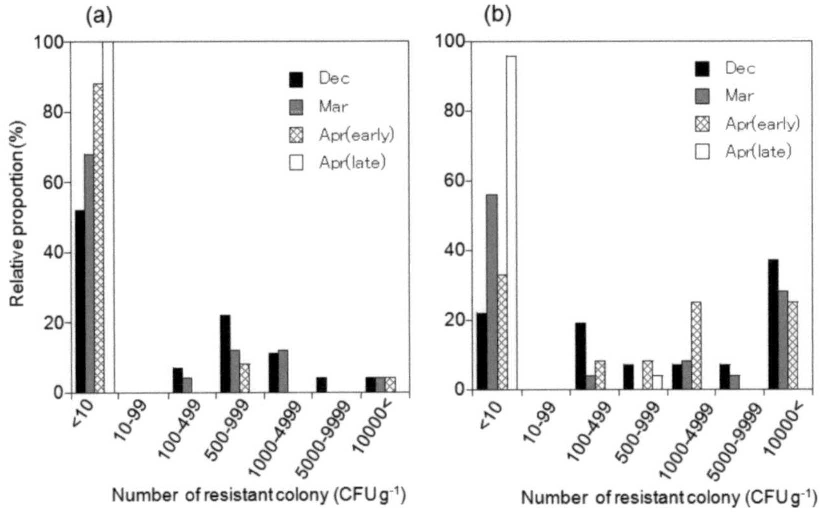

Figure 6.
Frequency distribution of the number of colistin-resistant E. coli (a) and coliform (b) in fecal samples of Eurasian wigeon. Fecal samples were taken around Ai River.

antibiotic used as the last resort for carbapenem-resistant Enterobacteriaceae bacteria at the World Health Organization. Thus, the succession of number of colistin-resistant *E. coli* and coliform in Eurasian wigeon was investigated in this study.

The frequency distribution of the number of colistin-resistant *E. coli* per gram of the sample is shown for each count range in **Figure 6**. In December 2017, the proportion of fecal samples below the detection limit was approximately 50%, and >100 CFU/g was about 50%. In March, the proportion gradually increased to approximately 70%, about 90% at the beginning of April, and 100% at the end of April 2018. For the colistin-resistant coliform, it was also found that the proportion of samples below the detection limit increased gradually while spending a winter in Japan. These results suggest that colistin-resistant *E. coli* and coliform may have been carried over to Japan after ingestion by the Eurasian wigeon in the northern area.

7. DNA barcoding and diet

DNA barcoding is a technique that allows identification of species by using a short nucleotide sequence of a specific gene region. By using a gene region that reflects the difference in species as a standard DNA barcode, it has become possible to specify species. This method can be used to identify species of plants, animals, and fungi. It also helps to discover new varieties by combining with other information.

For animals, about 650 bases in length at the 5′ end of the cytochrome C oxidase subunit I (*COI*) gene on the mitochondrial genome is regarded as a standard barcode region, but the primer used differs, depending on the research project [38, 39]. The reason that the *COI* of mitochondria was selected as a standard barcode region of DNA barcoding of animals is that the universal primers are available to cover most taxa of the animal kingdom and the region contains several mutations at the species level. With a few exceptions, the cells of all eukaryotic species contain the mitochondria. The mitochondrial genome comprises a double-stranded DNA molecule of approximately 16-kb length, accounting for 1–2% of the total DNA in mammalian cells.

Target gene	Name	Nucleotide sequence (5′-3′)
COI	LCO1490	GGTCAACAAATCATAAAGATATTGG
	HCO2198	TAAACTTCAGGGTGACCAAAAAATCA
matK	3F_KIM f	CGTACAGTACTTTTGTGTTTACGAG
	1R_KIM r	ACCCAGTCCATCTGGAAATCTTGGTTC
rbcL	rbcLa_F	ATGTCACCACAAACAGAGACTAAAGC
	rbcLa_R	GTAAAATCAAGTCCACCRCG
12S	M13U12S-F	TGTAAAACGACGGCCAGTCAAACTGGGATTAGATACCC
	M13U12S-R	CAGGAAACAGCTATGACCGAGGGTGACGGGCGGTGTGT
16S	M13U16S-F	TGTAAAACGACGGCCAGTACCGTGCAAAGGTAGCATAAT
	M13U16S-R	CAGGAAACAGCTATGACCTCCGGTCTGAACTCAGATCAC

*Underlined portions are M13 tag sequences [39].

Table 2.
Representative primers used for DNA barcoding.

On the other hand, for plants, a few mitochondrial interspecific mutations and *COI* cannot be used for species-level identification. Therefore, researchers propose the use of the chloroplast DNA region in plants, with *rbcL* and *matK* as standard barcode regions of terrestrial plants [40].

For analyzing the intestinal contents of various types of wildlife, the DNA barcoding method described above can be used in combination with NGS [41, 42]. Representative primers used for DNA barcoding were shown in **Table 2**. Various primer sets are available from the following site (http://www.boldsystems.org/index.php/Public_Primer_PrimerSearch).

8. Migratory bird diet

We focused on the DNA barcoding method to examine the relationship between diet and intestinal microbial communities in migratory birds. It can be hypothesized that diet—based on foods available in the area—has an influence, leading to change in the intestinal bacterial community of Eurasian wigeon and the number of colistin-resistant *E. coli* while spending a winter in Japan. Therefore, we performed NGS analysis on the intestinal contents of the Eurasian wigeon using eukaryotic *COI* gene as the target (**Figure 7**).

The *Orthocladiinae* sp. belonging to Arthropoda were found to be abundant at any time period, although there were differences in their proportion depending on the season. In December, *Demodex folliculorum* and *Orthocladiinae* sp., which belonged to Arthropoda, and *Adineta vaga*, which belonged to Rotifera, were abundant. Rhodophyta and Ochrophyta were observed only in the February samples. In the April samples, *Cricotopus annulator,* which belonged to Arthropoda, and *Adineta vaga*, which belonged to Rotifera, were abundant.

Pearson's correlation coefficients did not show any strong relationships between the number of both colistin-resistant *E. coli* and coliform and the diet. The abundance of Actinobacteria and Bacilli showed a significant but week correlation with abundance of *Scaptomyza montana* ($r = 0.340$, $P < 0.05$ and $r = 0.417$, $P < 0.01$, respectively). The abundance of Clostridia was negatively related with *Orthocladiinae* sp. ($r = -0.372$, $P < 0.05$) and *Scaptomyza montana* ($r = -0.327$, $P < 0.05$) and positively related with abundance of *Talaromyces pinophilus* ($r = 0.409$, $P < 0.01$).

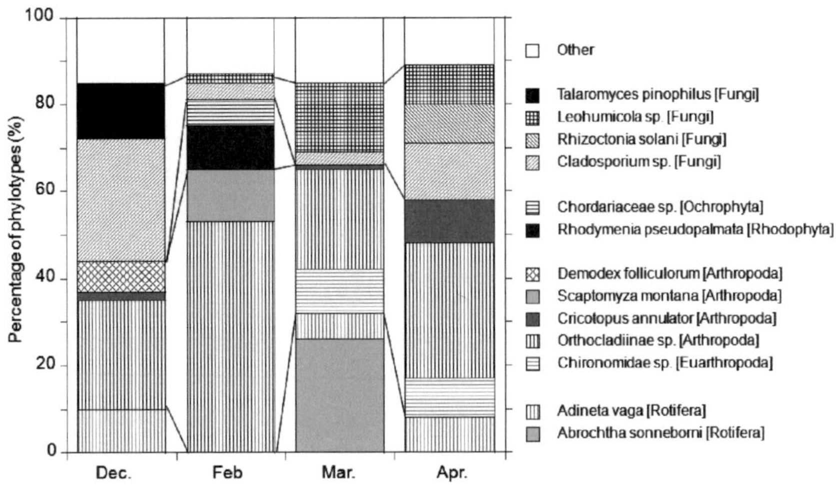

Figure 7.
Relative proportions of phylotypes of diet in the Eurasian wigeon.

The abundance of Gammaproteobacteria was positively related with abundance of *Abrochtha sonneborni* ($r = 0.354$ $P < 0.05$) and *Orthocladiinae* sp. ($r = 0.393$, $P < 0.05$, respectively). All correlation coefficients were not strong, and thus other parameters or multiple parameters might affect the intestinal microbial communities.

Although the Eurasian wigeon mainly ingests terrestrial plants, *COI* sequences of terrestrial plants were less detected in this study. For organisms that feed on plants, it seems better to use the sequences of *rbcL* and *matK* on chloroplast DNA. In order to investigate the relationship between the intestinal microbial community and diet in wild birds, it is necessary to select appropriate primers suitable for particular types of living organisms that are consumed as part of the diet.

9. Bacterial community composition in migratory and nonmigratory birds

In order to clarify the intestinal bacterial communities among different avian species, those of seagulls, Eurasian wigeon, and barn swallow were compared at their class levels (**Figure 8**). For this research, feces of the European herring gull and Slaty-backed gull were collected in Hokkaido (northern Japan). Slaty-backed gull (*Larus schistisagus*) is a large gull, measuring about 60 cm in length, breeding around northern Japan in the summer season and traveling to the south of mainland Japan and South Korea in the winter season. The global population was estimated to be 25,000–1,000,000 individuals, while the national population was estimated to include >1000 wintering individuals in Japan [43]. The European herring gull (*Larus argentatus*) is a large seagull with a total body length of approximately 60 cm, breeding in eastern Siberia in the summer season and traveling to Japan or to more southern areas in the winter season. These seagulls mainly eat fish, but they also eat crustaceans, insects, and other bird eggs. The global population was estimated to be 2,060,000–2,430,000 individuals, but the population was estimated to be decreasing at a rate of approximately 30% in 39 years [44].

Although the intestinal bacterial community of the Eurasian wigeon was dominated by Clostridia and Bacteroidia, those of the seagulls were dominated by Gammaproteobacteria. **Figure 8** shows the results of PCA, comparing the similarities between the intestinal bacterial communities of the migratory birds with other birds

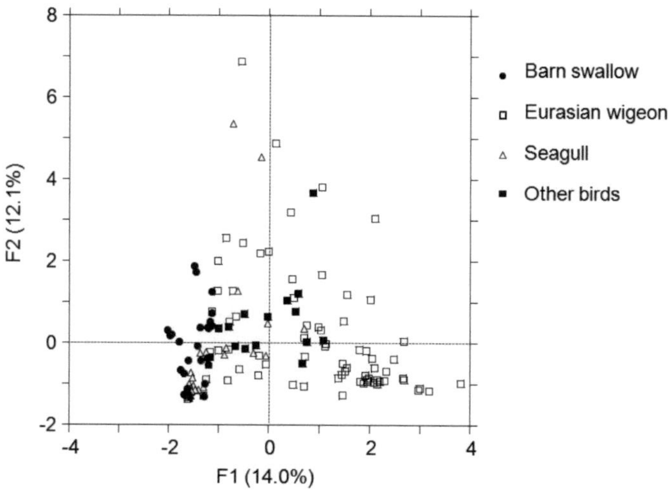

Figure 8.
Principal component analysis of class abundance data from migratory birds and nonmigratory birds.

registered on the public database. It is highly likely that migratory birds may eat different foods; therefore, differences across individuals were large as compared to those in poultry. However, as compared with other birds, individual intestinal microbiota from the barn swallow was relatively similar. In particular, intestinal bacterial composition from the Eurasian wigeon (□) collected from different seasons (December and April) was found to be highly diversified. The extent of the difference in them surpassed the extent of the difference among other birds. It may be reasonable that each of the intestinal bacterial communities was formed by the food consumed, be it an insect meal, an herbivorous meal, an omnivorous meal, or a carnivorous meal.

10. Conclusion

The use of culture-independent methods for studying wild bird-associated microbial communities could have been shown to be beneficial in the expansion of our current knowledge. The NGS targeting the 16S rRNA gene allows a comprehensive clarification of the bacterial communities, their succession while spending a winter or breeding, and their associated movement with migratory birds. The application of NGS is expected to improve our understanding of the overview of not only bacterial communities but also organisms ingested as part of the diet in wild birds. Narrowing down the target organisms using NGS will enable us to identify unknown pathogens or reveal the potential migration status of known pathogens that have escaped noticed so far due to methodological constraints. In addition, the relationship between intestinal microbial communities and diet of living organisms needs to be studied in greater detail.

Investigation of community composition in parallel with functional investigations (e.g., drug resistance) is expected to improve our understanding of the mechanisms by which multidrug-resistant bacteria spread around the world. Addressing the current implications of birds as potential vectors of antibiotic-resistant bacteria is of great interest. Analysis of the indigenous bacterial flora of migratory birds may highlight the importance of human hygiene and the environmental significance of the transfer of antibiotic-resistant bacteria associated with natural bird migratory patterns. When wild birds act as vectors of diseases, it is important to identify the true source of infectious organisms. NGS, as a culture-independent method, facilitates further understanding of the complexities and interactions of the genera inherently associated with birds, such as sputum, feces, and feather, as well as of those acquired from the wintering or breeding environment.

Acknowledgements

This work was supported by the JSPS Grant-in-Aid for Scientific Research (C) (18 K06435) and Grant-in-Aid for Scientific Research on Innovative Areas (15H05946). The author thanks Emi Kishino, Kiyo Horikawa, Saki Tsujita, and Haruka Takehashi for technical assistance during this work.

Conflict of interest

The author has no conflicts of interest directly relevant to the content of this article.

Author details

Takehiko Kenzaka
Faculty of Pharmacy, Osaka Ohtani University, Tondabayashi, Japan

*Address all correspondence to: kenzat@osaka-ohtani.ac.jp

References

[1] Mardis E. Next-generation DNA sequencing method. Annual Review of Genomics and Human Genetics. 2008;**9**:387-402. DOI: 10.1146/annurev.genom.9.081307.16435

[2] Morozova O, Marra MA. Applications of next-generation sequencing technologies in functional genomics. Genomics. 2008;**92**(5):255-264. DOI: 10.1016/j.ygeno.2008.07.001

[3] McGuire AL, Colgrove J, Whitney SN, Diaz CM, Bustillos D, Versalovic J. Ethical, legal, and social considerations in conducting the Human Microbiome Project. Genome Research. 2008;**18**(12):1861-1864. DOI: 10.1101/gr.081653.108

[4] Lloyd-Price J, Mahurkar A, Rahnavard G, Crabtree J, Orvis J, Hall AB, et al. Strains, functions and dynamics in the expanded Human Microbiome Project. Nature. 2017;**550**(7674):61-66. DOI: 10.1038/nature23889

[5] Jovel J, Patterson J, Wang W, Hotte N, O'Keefe S, Mitchel T, et al. Characterization of the gut microbiome using 16S or shotgun metagenomics. Frontiers in Microbiology. 2016;7:459. DOI: 10.3389/fmicb.2016.00459

[6] Benskin CM, Wilson K, Jones K, Hartley IR. Bacterial pathogens in wild birds: A review of the frequency and effects of infection. Biological Reviews of the Cambridge Philosophical Society. 2009;**84**(3):349-373. DOI: 10.1111/j.1469-185X.2008.00076.x

[7] Anderson JF, Andreadis TG, Vossbrinck CR, Tirrell S, Wakem EM, French RA, et al. Isolation of West Nile virus from mosquitoes, crows, and a Cooper's hawk in Connecticut. Science. 1999;**286**(5448):2331-2333. DOI: 10.1126/science.286.5448.2331

[8] Hurt AC, Vijaykrishna D, Butler J, Baas C, Maurer-Stroh S, Silva-de-la-Fuente MC, et al. Detection of evolutionarily distinct avian influenza a viruses in antarctica. MBio. 2014;**5**(3):e01098-e01014. DOI: 10.1128/mBio.01098-14

[9] Plutzer J, Tomor B. The role of aquatic birds in the environmental dissemination of human pathogenic *Giardia duodenalis* cysts and *Cryptosporidium* oocysts in Hungary. Parasitology International. 2009;**58**(3):227-231. DOI: 10.1016/j.parint.2009.05.004

[10] Petersen KD, Christensen JP, Permin A, Bisgaard M. Virulence of *Pasteurella multocida* subsp. multocida isolated from outbreaks of fowl cholera in wild birds for domestic poultry and game birds. Avian Pathology. 2001;**30**(1):27-31. DOI: 10.1080/03079450020023168

[11] Abulreesh HH, Goulder R, Scott GW. Wild birds and human pathogens in the context of ringing and migration. Ringing & Migration. 2007;**23**(4):193-200. DOI: 10.1080/03078698.2007.9674363

[12] Hubálek Z. An annotated checklist of pathogenic microorganisms associated with migratory birds. Journal of Wildlife Diseases. 2004;**40**(4):639-659. DOI: 10.7589/0090-3558-40.4.639

[13] Bonnedahl J, Järhult JD. Antibiotic resistance in wild birds. Upsala Journal of Medical Sciences. 2014;**119**(2):113-116. DOI: 10.3109/03009734.2014.905663

[14] Wang W, Zheng S, Sharshov K, Sun H, Yang F, Wang X, et al. Metagenomic profiling of gut microbial communities in both wild and artificially reared Bar-headed goose (*Anser indicus*). Microbiology. 2017;**6**(2):e00429. DOI: 10.1002/mbo3.429

[15] Ryu H, Grond K, Verheijen B, Elk M, Buehler DM, Santo Domingo JW. Intestinal microbiota and species diversity of *Campylobacter* and *Helicobacter* spp. in migrating shorebirds in Delaware Bay. Applied and Environmental Microbiology. 2014;**80**(6):1838-1847. DOI: 10.1128/AEM.03793-13

[16] Kenzaka T, Fujimitsu T, Kataoka K, Tani K. Intestinal microbiota in migrating Eurasian wigeon around Lake Biwa. Journal of Antibacterial and Antifungal Agents. 2018;**46**(3):101-104

[17] Kenzaka T, Kataoka K, Fujimitsu T, Tani K. Intestinal microbiota in migrating barn swallows around Osaka. Yakugaku Zasshi. 2018;**138**(1):117-122. DOI: 10.1248/yakushi.17-00148

[18] Kenzaka T, Tani K. Draft genome sequence of multidrug-resistant *Stenotrophomonas pavanii* BWK1, isolated from *Mareca penelop*e feces. Genome Announcements. 2018;**6**(12):e00187-e00118. DOI: 10.1128/genomeA.00187-18

[19] Kenzaka T, Tani K. Public health implications of intestinal microbiota in migratory birds. In: Kumavath RN, editor. Metagenomics for Gut Microbes. London: IntechOpen; 2018. pp. 35-51. DOI: 10.5772/intechopen.72456

[20] Novais RC, Thorstenson YR. The evolution of Pyrosequencing(R) for microbiology: From genes to genomes. Journal of Microbiological Methods. 2010;**86**(1):1-7. DOI: 10.1016/j.mimet.2011.04.006

[21] Xu HS, Roberts N, Singleton FL, Attwell RW, Grimes DJ, Colwell RR. Survival and viability of nonculturable *Escherichia coli* and *Vibrio cholerae* in the estuarine and marine environment. Microbial Ecology. 1982;**8**(4):313-323. DOI: 10.1007/BF02010671

[22] Yamaguchi N, Kenzaka T, Nasu M. Rapid in situ enumeration of physiologically active bacteria in river waters using fluorescent probes. Microbes and Environments. 1997;**12**(1):1-8. DOI: 10.1264/jsme2.12.1

[23] Kenzaka T, Yamaguchi N, Utrarachkij F, Suthienkul O, Nasu M. Rapid identification and enumeration of antibiotic resistant bacteria in urban canals by microcolony-fluorescence in situ hybridization. Journal of Health Science. 2006;**52**(6):703-710. DOI: 10.1248/jhs.52.703

[24] Kenzaka T, Yamaguchi N, Tani K, Nasu M. rRNA-targeted fluorescent in situ hybridization analysis of bacterial community structure in river water. Microbiology. 1998;**144**(8):2085-2093. DOI: 10.1099/00221287-144-8-2085

[25] Kenzaka T, Ishidoshiro A, Yamaguchi N, Tani K, Nasu M. rRNA sequence-based scanning electron microscopic detection of bacteria. Applied and Environmental Microbiology. 2005;**71**(9):5523-5531. DOI: 10.1128/AEM.71.9.5523-5531.2005

[26] Kenzaka T, Tani K, Nasu M. High-frequency phage-mediated gene transfer in freshwater environments determined at single-cell level. The ISME Journal. 2010;**4**(5):648-659. DOI: 10.1038/ismej.2009.145

[27] Iwamoto T, Tani K, Nakamura K, Suzuki Y, Kitagawa M, Eguchi M, et al. Monitoring impact of in situ biostimulation treatment on groundwater bacterial community by DGGE. FEMS Microbiology Ecology. 2000;**32**(2):129-141. DOI: 10.1111/j.1574-6941.2000.tb00707.x

[28] Muyzer G, de Waal EC, Uitterlinden AG. Profiling of complex microbial populations by denaturing gradient gel electrophoresis analysis of polymerase chain reaction-amplified

genes coding for 16S rRNA. Applied and Environmental Microbiology. 1993;**59**(3):695-700

[29] Lane DJ, Pace B, Olsen GJ, Stahl DA, Sogin ML, Pace NR. Rapid determination of 16S ribosomal RNA sequences for phylogenetic analyses. Proceedings of the National Academy of Sciences of the United States of America. 1985;**82**(20):6955-6959. DOI: 10.1073/pnas.82.20.6955

[30] Tringe SG, Hugenholtz P. A renaissance for the pioneering 16S rRNA gene. Current Opinion in Microbiology. 2008;**11**(5):442-446. DOI: 10.1016/j.mib.2008.09.011

[31] Baker GC, Smith JJ, Cowan DA. Review and re-analysis of domain-specific 16S primers. Journal of Microbiological Methods. 2003;**55**(3):541-555. DOI: 10.1016/j.mimet.2003.08.009

[32] Wang Y, Qian PY. Conservative fragments in bacterial 16S rRNA genes and primer design for 16S ribosomal DNA amplicons in metagenomic studies. PLoS One. 2009;**4**(10):e7401. DOI: 10.1371/journal.pone.0007401

[33] Yang B, Wang Y, Qian PY. Sensitivity and correlation of hypervariable regions in 16S rRNA genes in phylogenetic analysis. BMC Bioinformatics. 2016;**17**:135. DOI: 10.1186/s12859-016-0992-y

[34] Stoddard SF, Smith BJ, Hein R, Roller BRK, Schmidt TM. rrnDB: Improved tools for interpreting rRNA gene abundance in bacteria and archaea and a new foundation for future development. Nucleic Acids Research. 2015;**43**:D593-D598. DOI: 10.1093/nar/gku1201

[35] BirdLife International. 2016. *Hirundo rustica*. The IUCN Red List of Threatened Species 2016: e.T22712252A87461332. DOI: 10.2305/IUCN.UK.2016-3.RLTS.T22712252A87461332.en

[36] Kreisinger J, Kropáčková L, Petrželková A, Adámková M, Tomášek O, Martin J-F, et al. Temporal stability and the effect of transgenerational transfer on fecal microbiota structure in a long distance migratory bird. Frontiers in Microbiology. 2017;**8**:50. DOI: 10.3389/fmicb.2017.00050

[37] BirdLife International. 2017. Mareca penelope (amended version published in 2016). The IUCN Red List of Threatened Species 2017: e.T22680157A111892532. DOI: 10.2305/IUCN.UK.2017-1.RLTS.T22680157A111892532.en

[38] Hebert PD, Ratnasingham S, deWaard JR. Barcoding animal life: Cytochrome c oxidase subunit 1 divergences among closely related species. Proceedings of the Biological Sciences. 2003;**270**(Suppl 1):S96-S99. DOI: 10.1098/rsbl.2003.0025

[39] Yang L, Tan Z, Wang D, Xue L, Guan M-x, Huang T, et al. Species identification through mitochondrial rRNA genetic analysis. Scientific Reports. 2014;**4**:4089. DOI: 10.1038/srep04089

[40] CBOL Plant Working Group. A DNA barcode for land plants. Proceedings of the National Academy of Sciences of the United States of America. 2009;**106**(31):12794-12797. DOI: 10.1073/pnas.0905845106

[41] Barbosa A, Balagué V, Valera F, Martínez A, Benzal J, Motas M, et al. Age-related differences in the gastrointestinal microbiota of chinstrap penguins (*Pygoscelis antarctica*). PLoS One. 2016;**11**(4):e0153215. DOI: 10.1371/journal.pone.0153215

[42] Shehzad W, Riaz T, Nawaz MA, Miquel C, Poillot C, Shah SA, et al. Carnivore diet analysis based on next-generation sequencing: Application

to the leopard cat (*Prionailurus bengalensis*) in Pakistan. Molecular Ecology. 2012;**21**(8):1951-1965. DOI: 10.1111/j.1365-294X.2011.05424.x

[43] BirdLife International. 2016. *Larus schistisagus*. The IUCN Red List of Threatened Species 2016: e.T22694362A93450383. DOI: 10.2305/IUCN.UK.2016-3.RLTS.T22694362A93450383.en

[44] BirdLife International. 2016. *Larus argentatus*. The IUCN Red List of Threatened Species 2016: e.T62030608A89504806. DOI: 10.2305/IUCN.UK.2016-3.RLTS.T62030608A89504806.en

Chapter 8

Application of Metagenomics to Chilean Aquaculture

Mario Tello, Natalia Valdes, Rodrigo Vargas, Joselin Rojas, Mick Parra, Gonzalo Gajardo and Alex Gonzalez

Abstract

Aquaculture is a rapidly expanding food production sector, facing the challenge of growth both increasing its efficiency and reducing its negative impact on the environment. Metagenomics is an emerging tool in aquaculture that helps to understand the complex host (fish-shellfish)-microbiota-pathogen-environment relationship underlying disease outbreaks, monitoring the dynamics of microbial diversity in farmed animals subject to different environmental conditions or perturbations. As Chile takes an important share of world aquaculture market, this chapter reviews the actual and potential applications of metagenomics to support a sustainable expansion and diversification of Chilean aquaculture. The focus is on (i) the role and function of the gut microbiota in the proper immunostimulation and disease control and (ii) the role of metagenomics in monitoring environmental microbial biodiversity and dynamics in relation to disease persistence and ecosystem stability. We conclude that despite the importance of the aquaculture sector in Chile, the application of metagenomics to deal with disease control and ecosystem preservation is still an emerging field of study. Understanding host (fish-shellfish)-microbiota-pathogen-environment diversity of interactions in a more holistic view, i.e., the holobiome approach, could be key to develop rational strategies to improve productivity by increasing resistance to diseases and reducing the use of antibiotics and their negative environmental impact.

Keywords: metagenomics, Chilean aquaculture, microbiota, disease, environment

1. Introduction

Aquaculture, the so-called blue biotechnology of the future [1], is the fastest growing food production sector and should continue expanding and diversifying to fulfill the need for good-quality protein of the ~8.5 billion people that will populate the earth by 2030. Such ascending trajectory of aquaculture will also help to achieve some of the United Nations (UN) sustainable developmental goals (SDGs) [2] that seek a better, less unequal, and more sustainable future for humanity by 2030, specifically in relation to food security, improved nutrition, and poverty alleviation in rural communities in particular. Therefore, aquaculture growth and management of aquatic genetic resources should be done in a sustainable manner, a goal that requires maximizing ecosystem goods and services in line with the Blue Growth Initiative of the Food and Agriculture Organization (FAO) [2]. Aquaculture has expanded at a rate of 5.8% annually (2001–2016), totalizing ~ 80 million of tons in 2016, worth USD 231.6 billion (mainly finfish, crustacea, mollusk) [2]. Yet, the potential for expansion and diversification is immense considering the diversity of existing marine species [3]. For example, the total number of commercially farmed species increased in the last 10 years from 472 (2006) to 598 (2016) [2]. However, the sustainable aquaculture expansion is currently hampered by the impact of detrimental diseases causing high mortality and economic loss to the sector [4], together with the indiscriminate and inefficient use of antibiotics and chemicals to control them, which causes a negative ecosystem impact [5, 6]. To reevaluate this approach is the challenge ahead, and a good start is the ban imposed to the use of antibiotics by international organizations and consumers, which have stimulated the search for alternative microbial control strategies, like the use of probiotics [6]. Chile is a well-known and competitive salmon and trout producer. In fact, it's the world's second producer of salmon following Norway, and recently it has become a relevant mussel producer [2]. According to the FAO, in 2016 Chile ranked fourth among the world producers of marine and coastal finfish with 726.9 thousand tons (live weight) and the fourth marine mollusk producer (307.4 thousand tons [2]). The relevant aquaculture species are shown in **Figure 1**, with a total of 2162 centers distributed between the so-called Lake District (administrative region X); the northern part of Patagonia, where the salmon boom began; and the Magellan and Antarctic Region (XII). Region X is full of lakes that are intensively exploited as hatcheries for smolt production and also has protected coastal bays, fjords, and estuaries ideal for completing the marine phase of salmonid life cycle, not so far away from hatcheries. This region has 666 registered fish and 1171 mollusk centers. However, due to the ecosystem and disease consequences of the intensive salmon farming (high densities of fish per water volume) [7], the activity has moved to Region XI, with a total of 767 fish centers. In total (fish and mollusk), these centers harvested 1219.739 tons in 2016 [8].

One striking aspect of the farming of exotic salmonid species in Chile is the impressive expansion from the initial 80,000 tons harvested in the early 1990s to the 688,000 tons in 2004 and 900,000 tons in 2014 according to official statistics of the national service of fisheries, SERNAPESCA. In 2017 the production leveled at 791.103 tons with the following production figures by species (2017, tons): Atlantic salmon (*Salmo salar*) 582,350; coho salmon (*Oncorhynchus kisutch*), 134,235; and rainbow trout *(Oncorhynchus mykiss)* 74,518. Such successful story dates back to the last part of the twentieth century with the introduction of rainbow and brown trout

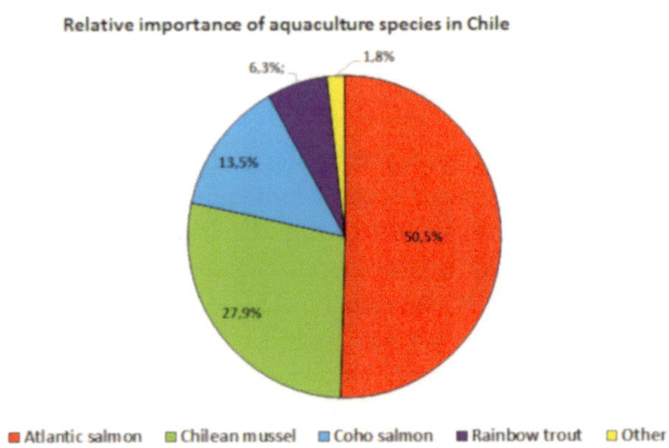

Figure 1.
Main species produced by Chilean aquaculture. According to production figures (2017), farming of salmon and trout is by far the main aquaculture activity in Chile, followed by the Chilean mussel (Mytilus chilensis). Others correspond to seaweeds and the Chilean scallop.

and other salmonids, initially for recreational fisheries and later for aquaculture [9]. However, the industry almost collapsed in 2006–2007 by the outbreak of the deadly virus responsible of the infectious salmon anemia (ISA), a problem that some anticipated due to the unlimited expansion of the industry in an ecosystem with limited carrying capacity [10]. This has raised serious concerns on the environmental standards of an industry making intensive use of coastal, estuaries, and lakes that are shared by multiple users and due to the high rate of antibiotic consumption (0.53 Kg per ton harvested in 2016), one of the highest in the world [7]. The interaction between host (fish), microbiota, and environmental microorganisms could be a key factor to develop rational strategies to improve the productivity by increasing the resistance to infection and reducing the use of antibiotics and the environmental impact of aquaculture. However, in the production scale, the intensive use of water and the utilities of the sector does not match with the lack of scientific research available to deal with the problems caused.

This chapter focuses on Chilean aquaculture to evaluate how metagenomics, a recently developed genomic subdiscipline, is actually contributing, or could potentially contribute, to develop more efficient aquaculture practices in relation to disease control and the environmental burden such practices have brought about. Metagenomics has made possible and cheaper the analysis of the complex genomes of microbial communities to unravel their diversity, dynamics, and functioning in different environments. The application of this tool to the aquaculture microcosm in particular has been reviewed by Martinez-Porchas and Vargas-Albores [11]. One of the benefits of metagenomics is to provide access to unculturable species, the vast majority of disease-related microbes in aquaculture whose diversity and function were unknown so far. We first provide an overview of Chilean aquaculture, the microbiological threats faced by the salmon farming industry, mainly the symbiotic and antagonist interrelationship between microbes and farmed animals. The focus is placed on how the gut microbiota of farmed and native species contribute to their fitness and overall performance in production-related traits like disease or stress resistance. Finally, we seek to evaluate the application of metagenomics to monitoring environmental biodiversity and microbial dynamics in a scenario of climate oscillations and other ecosystem perturbations such as the development of harmful algal blooms.

2. Microbiological threats facing Chilean aquaculture

As stated earlier, Chilean aquaculture is a successful industry. In fact, Chile is the first world producer of rainbow trout and coho salmon, is the second world producer of Atlantic salmon, and, recently, has become the world fourth producer of the Chilean mussel (*Mytilus chilensis*) [2, 8]. This level of production is achieved by the natural condition of the water present in the south of Chile and by the intensive farming strategy used. In fact, using less water surface, Chile competes to Norway in salmon production [12]. As consequence, Chilean aquaculture has faced several sanitary problems, related with outbreaks of viruses such as infectious pancreatic necrosis virus (IPNV) [13] and infectious salmon anemia virus (ISAV) [14]; bacterial pathogens such as *Piscirickettsia salmonis* [15], *Flavobacterium psychrophilum* [16], and *Renibacterium salmoninarum* [17]; and parasites such as *Caligus rogercresseyi* [15]. Diseases caused by viruses are one of the great challenges of Chilean salmon farming industry; the first virus being identified in Chile was IPNV, which is highly persistent and causes severe mortality. However, the introduction of individuals with QTL related to the resistance to the disease and the administration of vaccines has partially reduced mortalities in Atlantic salmon [15, 18, 19]. Due to its prevalence and because IPNV has been found in healthy fishes, it is now considered endemic in the Chilean Coast [13]. The ISAV, an orthomyxovirus related with influenza, caused several outbreaks between 2007 and 2008 forcing to close about 90% of the Atlantic salmon hatcheries located in Region X [14, 20]. To say the least, the industry almost collapsed. New outbreaks of ISAV have been detected in the last years; however, its pathogenicity is not comparable to the 2007–2008 outbreaks [15]. Recently, piscine reovirus (PRV) has been found in some Atlantic salmon with the so-called heart and skeletal muscle inflammation (HSMI) and also rainbow trout [21], although this virus has not been related to outbreaks with important losses [22].

Regarding bacterial pathogens, *P. salmonis* accounts for about 15% of production loss of post-smolt Atlantic salmon [15], while *F. psychrophilum* produces mortalities between 20 and 80% in rainbow trout and Atlantic salmon during the freshwater phase [23]. *R. salmoninarum* is the etiological agent of the bacterial kidney disease (BKD) and is responsible of the systemic and chronic infection in rainbow trout, Atlantic salmon, and coho salmon, with about 40% of prevalence in the latter species [24]. Several types of vaccines have been produced to control viral- and bacterial-caused diseases that, however, have proved inefficient as disease outbreaks continue to appear. This is the reason why antibiotics are still the main therapeutic strategy against bacterial pathogens. More than 186 tons were used only in the first semester of 2018 [25, 26].

Regarding farmed mussels (*Mytilus chilensis*) in Chile, there are no serious disease outbreaks reported so far, as it has occurred in major producer centers in Spain where congeneric species are intensively cultivated (*M. galloprovincialis* and *M. edulis*). Most of the registered mussel centers [8] located in the De Los Lagos Region (Lake District), southern Chile, cultivate mussels from ropes hanging from long lines. Spats are obtained either by the existing natural beds or by natural larva settlement. A recent review on bacteria associated to mollusk farming concentrated in the Chilean scallop (*Argopecten purpuratus*), a species cultivated in the north that showed potential aquaculture relevance prior to mussels [27]. This study also reports bacteria with probiotic effect on the pathogens listed. More recently, Lohrmann et al. [28] provided baseline information on the symbionts and other conditions of cultivated mussels from the Lake District, such as parasites (protozoan *Marteilia* sp., coccidian, and gregarines), intestinal copepods, castrating trematodes (*Proctoeces* sp.), intracellular bacteria in gills and digestive glands, ciliates on gills, microsporidian and metazoan parasites, and other conditions.

Currently, the factors that preclude bacterial and viral outbreaks remain unknown. With no doubt, the intensive culture strategy used by the industry accounts for disease prevalence and economic losses. But still, the cost–benefit ratio is extremely positive for the industry. Very likely, several unknown factors affect the host-pathogen-environment relationship. While the physicochemical factors enhancing disease outbreaks are relatively easy to determine, either by direct or satellite monitoring, the analysis of the microbiological factor is still biased to cultivable bacteria (see below), which means that more than 99% of the all microorganism present in the environment are left out. Metagenomics provides a more holistic view of the microbial ecosystem and their interactions, and so it is expected to shed light on this complex problem for the industry, for the environment, and also to humans that eat fish with antibiotic traces.

3. Microbiota of cultivated species in Chile

3.1 Microbiota

Multicellular eukaryotes (plants and animals) have traditionally been classified as highly complex organisms independent of the community of commensal microorganisms that colonize them [29, 30]. This community of microorganisms is known as microbiota and represents between 50 and 90% of an individual's cells in pluricellular forms of life [31]. Our initial failure to appreciate its importance is derived from biases arising from analyzing only cultivable microorganisms, which represent less than 1% of the diversity in a determined environment [32]. The development of non-culturing methods for microbial identification like PCR amplification of rRNA genes (16S or 18S) [33] and subsequent DGGE/TGGE analysis [34] and the development of massive sequencing techniques [35] allowed to broaden our knowledge, making possible to assess the complex community of microorganisms colonizing animals and plants such as bacteria, archaea, yeasts, and fungi. In humans, the gut microbiota is now considered as a complex endocrine organ that has coevolved with us through time, including cultural evolution [36–38], secreting several molecules that modulate human physiology [39]. The intimate and indissoluble relationship between microbiota and its host led to redefining the term organism and the emergence of the concept holobiont, which is used to define a community composed of host and hosted microbiota [40].

A healthy host has a stable microbiota which is altered when the metabolism or behavior of the host changes. In turn, changes in the microbiota composition caused by imbalances in the microorganisms that compose it (dysbiosis) can also produce metabolic changes in the host [41]. To date, there is abundant evidence showing that microbiota from mammalian participates directly in four processes: (a) protection against pathogens [42], (b) behavior [43], (c) energy balance [44], and (d) stimulation and maturation of the immune system [45, 46].

Much less is known about the characteristics and role of microorganisms that normally colonize the Atlantic salmon or rainbow trout and other teleost fishes [47]. Despite this, there is evidence showing a functional similarity between the roles of commensal microorganisms of salmonids and mammals [48]. In both cases, a complex community is established at the mucosal level that changes according to diet [49–56], temperature [57], season [57, 58], geographical location [59], culture condition [60–63], genetic [64], and stage of growth [58, 59, 65–67]. Microbiota composition also varies depending on the mucosal surface and epithelial location [68, 69]. High-resolution maps using next-generation sequencing (NGS) have identified around 950 operational taxonomic units (OTUs) in Atlantic salmon, with a slightly higher number in the skin [65, 68]. In the gastrointestinal tract (GT), these OTUs belong mainly to the phyla *Proteobacteria*, *Firmicutes*, and *Tenericutes* [55], while in the skin the main phylum is *Proteobacteria* [65]. As expected, the exposure to antibiotics such as oxytetracycline produces profound changes in culturable [70] and non-cultivable microbiota (Tello unpublished).

3.2 Function of the microbiota in salmonids

Although the function of microbiota in the physiology of salmonids has not been deeply studied, some evidence in Atlantic salmon and rainbow trout and other teleost fishes shows that it may play a similar role as described in mammals [48]:

a. *Antagonist against pathogens*: In salmonids and other fishes, the mucosa covers the entire organism; thus, it is estimated that microbiota colonizing it forms a main defense line against pathogens, by mechanisms that include competition for space or nutrients [71], interruption of communication signals, and production of inhibitory and antimicrobial substances, such as reported in mammalian [72]. However, few studies have focused on understanding the role of microbiota in salmonids. Some of them, performed in rainbow trout, show that bacteria from microbiota are able to protect against bacterial and fungal pathogens specially in egg or larval stages when the immune system is not fully developed [69, 73, 74]. On the other hand, some probiotic bacteria that have direct antagonism against salmonid pathogens confer protection in challenge assays, for example, *Clostridium butyricum* protects rainbow trout against infection with *Vibrio* sp. [75].

b. *Behavior*: Conditions that elicit stress in Atlantic salmon, rainbow trout, and zebra fish cause significant changes in the microbiota [76, 77]. This is probably due to increased mucosal secretion. Although it is not yet known if the commensal microbiota of rainbow trout or Atlantic salmon can modulate behavior, the administration of *Lactobacillus rhamnosus* IMC 501 significantly alters social and explorative behavior in zebra fish [78]. Moreover, in *Solea senegalensis*, *Shewanella putrefaciens* Pdp11 prevents mortality caused by increased population density [79], and in other fishes the administration of bacterial probiotics reduces levels of cortisol, a molecule with immunosuppressive activity that is secreted under stress conditions [80]. It remains to be determined if this protection is due to reduced sensitivity to the stress caused by high population density, increased resistance to opportunistic pathogens through competition or immunostimulation, or both.

c. *Energy balance*: In Atlantic salmon or rainbow trout, there is some evidence showing the microbiota is implicated in harvesting energy. The administration of probiotics for 10–12 weeks increases the rate of conversion in rainbow trout [81]. In zebra fish, the microbiota stimulates genes related to nutritional metabolism [82], increasing fat uptake and storage in adipose tissue [83], and also stimulates maturation of the gastrointestinal tract by increasing the production of digestive enzymes [84].

d. *Stimulation and maturation of the immune system*: In mammalians and zebra fish, the role of microbiota in the development of a proper immune response, especially in the maturation of innate immune response, is well known. In the gnotobiotic zebra fish model, the treatment with nonabsorbable antibiotics to deplete normal microbiota improves epithelial proliferation [85]; the innate immune response [82]; the expression of immunoglobulin M gene [86], neutrophils, macrophages, and phagocytosis [87–89]; and its composition also determined by B- and T-cell receptors [90]. In other fishes, the administration of probiotics helps to improve phagocytosis and the alternative complement pathway [91].

In Atlantic salmon or rainbow trout, it is unknown if normal microbiota modifies the immune response, but some indirect evidence suggests:

1. Gut inflammation occurs as a consequence of changes in diet that alter the microbiota [92, 93]. It remains to be determined if inflammation is caused by the change in microbiota or not.

2. Probiotics improve the immune function and innate immune response [75, 94, 95].

3. Most of the elements that affect the host-microbiota relationship in mammalians and zebra fish are also observed in salmonids (see below).

As mentioned earlier, evidence shows that microbiota should play several roles in the salmonid physiology. However, functional association between metagenomic characterization and physiological parameters is an amazing challenge poorly explored. Likewise, the identification of the microorganism or consortium unique to fishes that could be administrated as probiotics to improve the performance of salmonid aquaculture is, therefore, a challenge ahead.

3.3 Effect of antibiotics

In agricultural systems, the use of chemical products, especially antibiotics, has become widespread as a form of disease prevention and treatment [96]. While, in cattle breeding, antibiotics are preferably used as growth promoters [97], in aquaculture these properties have not been convincingly demonstrated. Historically, due to its intensive practice, the Chilean salmon industry is identified as one that uses more antibiotics per ton of harvested product [7]. In 2016, the amount of antibiotic used in the Chilean salmon industry reached 382.5 tons. The administration of antibiotics in aquaculture farms is mainly done through food, the remaining of which accumulates in the environment together with excretions [98]. Such accumulation of antibiotics in the marine sediments can persist for months, acting as a selection pressure that favors the establishment of resistant microorganisms that alter the endemic microbiota and the natural biogeochemical processes [99]. In Chile, this phenomenon is of great importance, since most of the salmon production is concentrated in the south of Patagonia, an area of high biological diversity.

The global trend in salmon production has been to reduce the use of chemical products to comply with biosecurity and animal welfare policies. In Chile, however, the use of antibiotics has continued to increase, accumulating an average of 343.4 tons of antibiotics per year (period 2005–2016) [100], 95% of which is used in marine fish farms. During this period, a high rate of infection by the intracellular bacterium *Piscirickettsia salmonis*, which causes the salmonid rickettsial syndrome (SRS), was observed, for which there is no effective vaccine or antibiotic treatment [101].

The main antibiotics used in Chile correspond to florfenicol and oxytetracycline. According to the national service of fishing and aquaculture from Chile (SERNAPESCA) in 2016, florfenicol and oxytetracycline represented 82.5 and 16.8%, respectively, of all the antibiotics used in Chilean aquaculture [100]. Both are broad-spectrum antibiotics used to combat infections of aquatic pathogens such as *Aeromonas salmonicida*, *Aeromonas hydrophila*, and *Yersinia ruckeri*, among others [102]. The predominance of florfenicol in recent years is mainly due to the fact that this antibiotic is the main agent against *P. salmonis*.

Pathogens cause immense economic losses that also have social impact. The crisis of the ISA virus in 2007 represented US$ 600 million and 16,000 jobs lost. The infection by *P. salmonis* caused 70% of the mortality of Atlantic salmon and rainbow trout in recent years, amounting to US$ 450 million per year, including vaccination, antibiotics, and other measures to mitigate the disease [103].

Although preventive measures have been implemented with the mandatory use of vaccines, the results have not been as promising as in mammals. The reason behind would be the less developed immunological memory in salmonids than mammals [104]. A corollary of this is the continued massive use of antibiotics. As mentioned, a critical problem is the propagation in the environment of microorganisms with resistance genes. Antibiotic residues have been reported in the muscles of fish and can be transferred directly to humans if the fish is not cooked properly [105]. Studies using massive sequencing to detect antibiotic resistance genes from fish sediment identified that more than 90% had mobile genetic elements of high homology to human pathogens [106]. This confirms the high rate of genetic exchange, or horizontal transmission, of antibiotic resistance between microbes and fishes that in the end affects human health.

Some studies identify the presence of antibiotic resistance genes in isolates from areas where salmonids are cultivated [107]; however, in-depth studies on the impact of antibiotics on the composition of microbial communities in farmed animals are still lacking. This field is open for metagenomics, especially for environmental DNA monitoring, in order to evaluate the impacts on the native microbial communities and their dynamics in places where salmon cages are located. It is particularly important to establish how environmental microbial communities recover after antibiotic treatment or after cages have moved to another place to allow recovery of the site, according to the actual practice.

In relation to the effect of antibiotics on the salmonid microbiota, studies conducted with culturable bacteria indicate that oxytetracycline decreases bacterial diversity, facilitating the proliferation of opportunistic pathogenic bacteria [70]. Studies conducted in our laboratory using broad-spectrum antibiotics such as bacitracin/neomycin showed the bacterial load in the intestine is reduced 10 times, in particular, the population of *Proteobacteria*, which favors the increase of the phylum *Firmicutes*. At the level of genera, the predominance of *Pseudomonas* and *Aeromonas* is replaced by *Lactococcus*. The interesting data is that the microbial composition is not recovered after 15 days of antibiotic treatment, which suggests that changes in the microbiota could be irreversible, at least within a short window of time (Tello, unpublished data). At the functional level, we were able to see that antibiotics increase the diversity of genes related to general metabolic pathways (amino acid biosynthesis, secondary metabolites, enzyme synthesis, etc.) and antibiotic metabolisms.

In farming systems, the impact of antibiotics on the fish's normal microbiota and its effects on the long term is unknown. It remains to be determined if the dysbiosis induced by the antibiotics generates a favorable scenario for other pathogens or if it affects the immune response. Preliminary work from our laboratory indicates that the administration of bacitracin and neomycin induces the expression of the inflammatory immune response judging by the increase in the expression of interleukin IL-1b and the reduced amount of leukocytes in the immunological organs. Similar effects are observed when administering florfenicol, a broad-spectrum antibiotic widely used in salmon farming (**Figure 2**).

3.4 Effect of heavy metals

It is a well-known fact that the presence of metals in different environments generates a toxic effect on both the biota and microbiota. Metagenomic analyses show

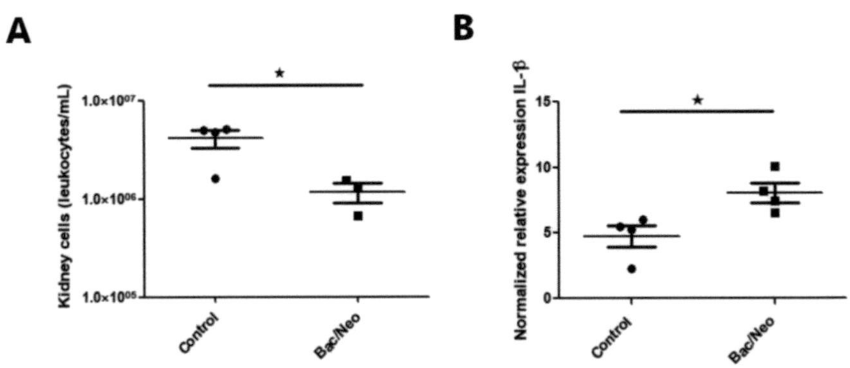

Figure 2.
Effects of bacitracin/neomycin on the immune system of Atlantic salmon. The figure shows the effects of bacitracin/neomycin administered by 14 days on the amount of leukocytes (A) and expression of IL-1b (B). Both measures were determined on the head kidney.

these effects are persistent in the gut microbiota and in the environment, principally in water bodies where the toxicity of these metals modulates the microbial community composition [108, 109]. Previous studies have demonstrated that environmental pollutants affect the gut microbiota even at low concentration [109]. The metagenomic analyses of the gut microbiota from *Cyprinus carpio* have made evident the prevalence (or selection) of bacterial genera containing genes associated to metal resistance as well as genes involved in heavy metal biotransformation pathways that tend to attenuate their toxicity in fishes exposed to heavy metals [110, 111].

The Chilean aquaculture industry uses heavy metals such as copper due to its antimicrobial properties in, for example, antifouling paints that avoid the formation of bacterial biofilms which also are considered potential pathogen reservoirs [112, 113]. Another application of copper is for coating cages and recirculation systems (RAS) used in land-based salmon farming sites. On a less positive side, it has been observed that small variations in the concentration of copper in incoming freshwater from underground wells that feed a hatchery modify the fish behavior and reduce food intake, such as in cascade effect that ends up affecting the production. Such behavior impairment by copper seems to affect the nervous system of the fish [114]. Other studies in *Cyprinus carpio* show that low copper concentrations also affect the microbiota diversity and lipid metabolism [110].

Owing to the importance of recirculation systems in aquaculture facilities, it is imperative to understand how environmental pollutants affect the dynamic of the gut microbiota and health of reared fishes as well as the microbiota of the biofilter [115]. The metagenomic approach combined with metabolomic studies may help to understand the complex changes that occur at different levels in a hatchery exposed to environmental pollutants.

3.5 Potential use of microbiota of wild and farmed fishes in Chile

3.5.1 Metagenomics in the study of microbiota of wild and farmed fishes in Chile

Despite the importance of the gut microbiota in the fitness of wild and farmed animals, as seen for salmonids in Section 4, few studies have characterized the microbiota composition of farmed and wild fish and shellfish that are commonly consumed in Chile [107, 116–121]. Most of the best characterized microbiota corresponds to foreign species such as Atlantic salmon and rainbow trout, in agreement with their economic and social importance. With few exceptions, most of the microbiota reported in these species corresponds to studies done in Europe, Canada, the USA, and New Zealand. Even for rainbow trout, a species introduced in Chilean rivers more than 100 years ago [9], there are no studies characterizing and comparing the microbiota of feral rainbow trout living in Chilean rivers with the microbiota present in wild rainbow trout from regions where the species is autochthonous [122].

Since rainbow trout is a species considered “naturalized” in Chile, the obvious question would be if naturalization or adaptation to local conditions is associated with a particular microbiota composition. This is, perhaps, another dimension to consider in the study of the adaptation of rainbow trout in Chile. Given the wild north–south distribution of the species in Chile, it would be also necessary to understand if microbiota composition varies with latitude. The same sort of questions are pertinent to understand the microbiota composition of farmed trout and naturalized ones.

A similar situation happens with farmed Atlantic salmons which escaped from Chilean hatcheries. It is unknown if they show a microbiota composition similar or different to those present in the wild Atlantic salmon from Canada or Norwegian rivers or closer to the Atlantic salmon farmed in Chile. In the case of coho salmon, another heavily cultivated species in Chile, studies on the microbiota have been done using 16S rRNA PCR coupled to DGGE [123], which gives a sort of qualitative approach to the microbiota diversity. This study shows that stable microbiota is established after first feeding and comes mainly from bacteria located in eggs and water. A more accurate analysis of the genomes of the microbiota would be important to understand the function that a particular group of bacteria could play in the species adaptation to the aquaculture or wild environment.

In short, the study of microbiota of farmed salmonids, as well as or other species, and their naturalized or wild relatives should provide evidence on how the host-microbiota-environment relationships evolve [64, 124]. The microbiota from wild fishes (marine or freshwater) living in the Chilean territory could give us some clues to understand how these fishes have adapted to local conditions. Such studies should help to optimize its nutrition and protection against pathogens in the artificial environment under the environmental conditions present in Chile.

The intestinal microbiota of *Seriola lalandi* and *Paralichthys adspersus* (fine flounder) has been sequenced. Both species distributed in the southern Pacific Ocean have, respectively, actual or potential aquaculture interest in Chile. In both cases, bacteria belonging to the phylum *Proteobacteria* were the most abundant group in wild specimens, while under aquaculture conditions, members of the phylum *Firmicutes* predominated. Under aquaculture conditions, the fine flounder shows a reduction of *Actinobacteria*, a group known to produce antimicrobial compounds [117, 119]. Further, metagenomic characterization of the microbiota through the complete sequencing of all the microbial genomes is necessary to properly predict the function of the microorganisms that colonize wild or farmed fishes and so to have a better approach to the evolution of the host-microbiota relationship in *Seriola lalandi* and *Paralichthys adspersus*.

3.5.2 Potential uses of the metagenomics for identification of new probiotics

The term microbiota refers to a complex and dynamic ecosystem of microorganisms that colonize the exposed surfaces and epithelia of an organism [125]. The role of these microorganisms has been studied mainly in the gastrointestinal tract (GT), contributing greatly to the general welfare of the host, participating in the absorption of nutrients, functioning as a protective barrier against potential pathogens, and regulating the expression of genes involved in epithelial proliferation in addition to having a role in the stimulation of the immune system and the prevention of diseases [126]. It is for these reasons that the microbiota is a good source of probiotic potentials.

Probiotics are defined as live microbiological food supplements with beneficial effects for the animal host [127], which confer protection (antagonism) against pathogens, helping in the development of the immune system and providing nutritional benefits [128]. In the aquaculture sector, probiotics began to be used at the end of the 1980s, as a prophylactic method against pathogens, mainly due to their ability to stimulate the innate immune response, which is characterized by having a nonspecific mode of action against various microorganisms [129].

The microorganisms used as probiotics in the aquaculture can have different origins, microorganisms previously used in mammals (allobiotic, probiotic) or commensal microorganisms that colonize the GT and the mucous membranes of the fish (autochthonous probiotic). In the case of aquatic animals, probiotics of autochthonous origin have adaptive advantages against foreign microorganisms, since they are adapted to factors such as water temperature and salinity [91], a situation that allows them to compete adequately with the resident organisms of the GT [130], thus ensuring adequate colonization and persistence of the microorganism.

In the last three decades, several microorganisms with probiotic activity have been identified and characterized for the aquaculture sector [131]. These microorganisms include both Gram-positive and Gram-negative bacteria, bacteriophages, microalgae, and yeasts [132, 133]. Among the most used probiotic species in the production of salmonids are the genera *Lactobacillus*, *Bifidobacterium*, *Aeromonas*, *Plesiomonas*, *Bacteroides*, *Fusobacterium*, *Carnobacterium*, *Eubacterium*, *Bacillus*, *Enterococcus*, *Bacteroides*, *Clostridium*, *Agrobacterium*, *Pseudomonas*, *Brevibacterium*, *Microbacterium*, *Staphylococcus*, *Streptomyces*, *Micrococcus*, *Psychrobacter*, *Pediococcus*, *Saccharomyces*, *Debaryomyces*, *Alteromonas*, *Tetraselmis*, *Roseobacter*, *Weissella*, and *Aspergillus* [75].

The use of these and other microorganisms with probiotic activity has generated a reduction in the levels of antimicrobial compounds, particularly antibiotics, used in the salmon industry. In addition, there has been an improvement in the appetite and/or growth of farmed salmonids [134]. Many of these microorganisms have antagonistic activity in vivo in salmonids against pathogens such as *A. salmonicida*, *V. anguillarum*, *V. ordalii*, and *F. psychrophilum*.

Marine bacteria also have the potential to be used as probiotics. These bacteria have the ability to store the biodegradable polymer polyhydroxybutyrate (PHB), which exhibits the ability to neutralize pathogens in *Artemia*, fish, or shrimps. Such probiotic effect seems associated with the breakdown of PHB into monomers (short-chain fatty acids (SCFA)) in the gut of the target species; this breakdown changes pH and improves bacterial richness [135] or enhances immunological defense and provides energy to cells [6, 135–139]. Baruah et al. demonstrated that a commercial PHB source enhanced the survival of *Artemia* challenged with pathogen by triggering the expression of the heat shock protein, Hsp 70, which is associated with protective innate immune responses [135]. Another positive PHB effect has been reported in the European sea bass (*Dicentrarchus labrax*) in an experiment in which the diet of juveniles was supplemented with 2 and 5% PHB (w/w). Juveniles showed better growth performance correlated to a high bacterial richness in the gut [135]. Similar results were observed in the Siberian sturgeon (*Acipenser baerii*) fingerling, also with a diet supplemented with increasing amounts of PHB [137], as well as in shrimps [136, 139].

There are currently commercialized probiotics for use in aquaculture, such as Mycolactor Dry Probiotic®, which corresponds to a mixture of *Saccharomyces cerevisiae*, *Enterococcus faecium*, *Lactobacillus acidophilus*, *L. casei*, *L. plantarum*, and *L. brevis*; INVE Sanolife® MIC that includes a mixture of *Bacillus* strains (Biogen®), *Bacillus licheniformis* and *Bacillus subtilis*; and BACTOCELL® (*Pediococcus acidilactici*), the first probiotic approved by the European Union for use in aquaculture, as an additive in the feeding of salmonids [140].

In the case of mollusks, there is a history of a bacterial strain isolated from the gonads of Chilean scallops (*Argopecten purpuratus*) and characterized as *Alteromonas haloplanktis* which shows inhibitory activity in vitro against the known pathogens *V. ordalii*, *V. parahaemolyticus*, *V. anguillarum*, *V. alginolyticus*, and *A. hydrophila* [141]. The combination of *A. haloplanktis* and *Vibrio* strain 11 showed in vitro inhibition against *V. anguillarum*-protected scallop larvae in in vivo assays [142]. A recent example in the European blue mussel (*Mytilus edulis*) showed high poly-β-hydroxybutyrate levels regulating the immune response of mussels challenged with *Vibrio coralliilyticus* [143].

Other studies test the protective capacity of *Aeromonas media* A199 in vitro against other 89 strains of *Aeromonas* and *Vibrio*, in addition to preventing the death of oyster larvae (*Crassostrea gigas*) when challenged in vivo with *Vibrio tubiashii*. However, *A. media* A199 was not detected in the host after 4 days of the administration of the probiotic treatment, indicating that it would be necessary to administer the probiotic at regular intervals of time if a prolonged protective effect is required [144].

The functional relationship between the immune system of teleost and mammalians and innate and cellular response present in shellfish makes plausible that microbiota plays these roles in all cultured species. Characterization of the microbiota by a metagenomic approach has helped to identify microorganisms or consortia that can be used to improve the absorption of nutrients, have an antagonistic effect against bacterial pathogens, or can stimulate the innate and cellular response [145]. Metagenomics based on sequencing the 16 s rRNA associated to a host biological property or condition such as resistance to pathogens could help to identify bacteria or consortium with antimicrobial properties and look for ways to culture this bacteria to isolate potential probiotics. This analysis can also be complemented with a prediction of the metabolic pathway using the software PICRUSt [146]. It may also help to predict a particular condition of the fish if associated to a particular group of microorganisms with different metabolic properties, such as the production of vitamins, use of different carbon sources, or production of metabolites with immunomodulator properties such as SCFA or PUFA [118]. This analysis could be improved using metagenomics based on the complete sequence of the whole microbial DNA. This analysis, associated to ORF prediction, metabolic reconstruction, and prediction of secondary metabolites and antimicrobial peptides using antiSMASH [147], could help to improve the metabolic characterization of the microbiota associated to a particular condition and help to guide the identification of cultivable microorganisms that can be used as probiotics. Currently this approach began to be applied in the identification of potential probiotics to the aquaculture, for example, from eggs of Rainbow trout resistant to the infection with the fungus Saprolegna, was isolated a cultivable bacterium belonging to the genus *Actinomyces* that produce antifungal compounds and confer protection against this pathogen [73].

4. Importance of microbiota-host-environment interactions for the development of a sustainable aquaculture

4.1 Microbiota and immune system interaction in salmonids

Most of the current information about the host-microbiota communication comes from studies in mammalian models. These studies show that the interplay between the microbiota and the immune system is a sort of chemical cross talk [148] which involves from microbiota to host (a) the specific stimulation of host receptors by molecules from microbial organisms (bacteria, fungus, viruses, and archaea), (b) bioactive molecules secreted by the microbiota metabolism, and (c) stimuli of epigenetic mechanisms to control expression of immune genes and fate of immune cells. In salmonids the interplay between immune system and microbiota has not been studied yet. However, the immune system of salmonids shows the same elements that allow the host-microbiota communication in mammalians.

a. **Specific stimulation of host receptors by molecules from microbial organisms.** The immune system is able to recognize the pattern of some structural molecules from microorganisms either commensal or pathogens.

Characterized originally in pathogens, these pathogen-associated molecular patterns (PAMP), which are specific for each type of microorganism, are recognized by a family of receptors located in the immune cells denominated as pattern recognition receptor (PRR), in particular a subtype of them, namely, Toll-like receptors (TLR). This interaction appears to be key in stimulating NF-kβ, interferon-response factors, and the inflammasome, which, depending on the cell type, context, and microorganism involved, responds through the production of different inflammatory cytokines (e.g., interferon, IL-1β, IL-22), which can have systemic or local effects [149–153]. On the other side, commensal microbiota controls the immune response through the exposure or secretion of molecules that stimulate anti-inflammatory response. Polysaccharide A of *Bacteroides fragilis* can induce an anti-inflammatory response by increasing IL-10 production and the population of regulatory T lymphocytes and by decreasing the inflammatory response mediated by Th-17 [154, 155].

There is a greater diversity of TLR in salmonids than in mammals (20 versus 10) [156, 157], which suggests that in Atlantic salmon or rainbow trout, the cellular immune response mediated by TLR stimulation should be more complex than in mammals. This might be because fish live in an aqueous environment where they are exposed to a greater quantity and diversity of microorganisms that interact with a proportionally greater surface of mucosa. In addition, salmonids have a poorly developed immune response at the level of antibodies, strongly suggesting that in these organisms the cellular immune response and microbiota are the main barriers against pathogens. As in mammals, TLR expression is also stimulated by microbial infections [158]; however, how the pattern of TLR gene expression changes as consequences of variation in microbiota composition is unknown.

b. **Bioactive molecules secreted by the salmonid microbiota**. In mammals microbiota secretes several bioactive molecules able to modify the cell metabolism and immune response [159]; among them the molecules with the most significant impact are short-chain fatty acids (SCFA, formate, acetate, n-propionate, n-butyrate, and n-valerate). SCFA are generated by the anaerobic fermentative metabolism of bacteria that are part of the intestinal microbiota and, because they are hydrophobic, are absorbed by epithelial cells and rapidly disseminate throughout the organism causing effects in different organs [160]. The microbial SCFA best characterized is butyrate; this molecule induces the production of regulatory T lymphocytes beyond the thymus [161], stimulates microglia maturation and function and PMN lymphocyte activity [162, 163], and decreases the production of proinflammatory cytokines in macrophages (INF-γ, IL-1β, TNFα) [164]. Butyrate also decreases the proliferation and increases apoptosis of T CD4 lymphocytes [165, 166], increases the production of anti-inflammatory cytokines in dendritic cells (IL-10, IL-23), and decreases exposure of MHCII [167, 168]. Although in general the effect of butyrate and other SCFA is to promote anti-inflammatory responses, the exact role depends on the type of cell and SCFA. The generic effects of butyrate can be explained by its capacity to stimulate the free fatty acid receptors (GPR41, GPR43, and GRP109a), which in turn stimulate a cascade of phosphorylation by Gai to activate at ERK1/ERK2 MAP kinase. In mammalians, these receptors are expressed in immune cells, specially GPR43, which is highly expressed in macrophages/microglia, neutrophils, and monocytes [162, 164]. Bioinformatic assays performed in our laboratory indicated that Atlantic salmon genome encodes for 13 proteins which are homologous to the butyrate receptors present in mammalians. This expansion suggests an important role of this molecule in Atlantic salmon physiology [Tello et al. unpublished]; however, the pattern of expression of these genes is currently unknown in both the salmonid gut and its immune organs. Also it is unknown if butyrate is able to induce ERK1/ERK2 phosphorylation in salmonids.

Besides its interaction with its receptors, n-butyrate is also a strong inhibitor of histone deacetylase (HDAC), inducing chromatin remodeling and changes at an epigenetic level [169, 170]. HDAC is a highly conserved protein among different species. Human and Atlantic salmon HDAC shares a 97% of sequence identity, thus making highly plausible that HDAC from salmonids can also be inhibited by butyrate.

It is unknown if the salmonid microbiota produces butyrate or other bioactive molecules able to modulate the immune response; evidence from other fish suggest that butyrate is also a bioactive molecule in teleost. Butyrate is found in the intestinal tract of herbivorous and carnivorous fish [171, 172]. In *Sparus aurata* it increases intestinal microvilli and nutrient absorption [173] and in *Cyprinus carpio* increases the expression of shock protein-70 (HSP70), proinflammatory cytokines (IL-1β and TNF-α), and anti-inflammatory cytokines (transforming growth factor-β) [174]. The mechanism by which butyrate can induce these changes is currently unknown.

Preliminary experiments performed in our laboratory show that butyrate modifies the antiviral response in SHK-1 cells, in a mechanism that is independent of the expression of the putative butyrate receptors [Tello unpublished], suggesting that Atlantic salmon cells could be sensible to this microbial metabolite.

c. **Stimuli epigenetic mechanisms to control gene expression and fate of immune cells.** In mammals microbiota also controls the immune system through several epigenetic mechanisms: DNA methylation [175], histone modification [169, 170, 176, 177], and control of gene expression by noncoding RNA [178]. In most of cases, this control is achieved by metabolic products of the microbiota, such as butyrate or vitamin precursor [179]. In simple terms, DNA methylation of cytosine impedes transcription factor binding and favors the recruitment of methylated binding domain proteins, which in turn prevents the binding of transcription factors by inactivating the chromatin configuration around genes. Through changes in DNA methylation pattern, microbiota may control the proliferation of Treg [180] and the function of NK cells [181].

Changes in the histone modification pattern produced by inhibiting HDAC with SCFA stimulate changes in the chromatin structure increasing the expression of *foxP3*, which promotes the differentiation of T CD4+ lymphocytes in Treg lymphocytes, favoring the anti-inflammatory response [161]. In intestinal macrophages, SCFA reduces the production of proinflammatory mediators (cytokines) via HDAC inhibition [182]. Deleting histone deacetylase 3 in intestinal epithelial cells alters normal microbiota, changes the expression patterns of antimicrobial peptides, and increases inflammatory processes, suggesting that epigenetic control of the host by microbiota is a fundamental element in homeostasis maintenance [183]. Although the epigenetic regulation in Atlantic salmon or rainbow trout has been poorly studied, at genetic level, both species show a more complex DNA methylation and histone acetylation/deacetylation systems than mammalians, with several gene duplications [184], suggesting that this mechanism could also be implied in the host-microbiota communication.

Noncoding RNAs (miRNA, lncRNA, and snRNA) are a group of RNAs highly expressed in cells with several regulatory functions. Among them, microRNAs (miRNAs) are implicated in the cross talk between mammalian microbiota and its host immune system [178, 185, 186], while lncRNA are involved in the cross talk with gut epithelial cells [187]. Commensal microbiota is able to regulate the expression of several miRNAs that target genes involved in the inflammatory process, generating a tolerance state in the gut [188–190]. Among them, two important miRNAs that regulate the inflammation process are miRNA146 and miRNA155. miRNA146 expression is induced by low doses of LPS and acts as an anti-inflammatory regulator by targeting TNF receptor-associated factor 6 (TRAF6) and IL-1R-associated kinase 1 (IRAK1), which are involved in the NF-κB pathway. miRNA146 allows the establishment of postnatal intestinal microbiota in the newborn gut, preventing inappropriate inflammation. miRNA155 is induced by high concentrations of LPS and plays an opposite role stimulating the inflammatory process by targeting the negative regulator of the NF-κB pathway. Recent works using NGS (RNAseq) from different organs of Atlantic salmon identified 180 distinct mature miRNAs belonging to 106 families of miRNAs [191]. These miRNAs were deposited in the miRBase database (http://www.mirbase.org). This database currently contains 371 miRNAs from Atlantic salmon. Among the miRNAs identified in Atlantic salmon, orthologous of miRNA146, miRNA155, and other 15 of 27 miRNAs that participate in the microbiota-host communication in mammals were found. Currently it is unknown if the expression of these miRNAs changes according to the composition of Atlantic salmon microbiota.

The study of the mechanisms underlying the stimulation of immune system by microorganism that conform the microbiota is an open field for metagenomic studies searching for association between microorganism, consortia, or microbial metabolites and the proper immunological function. This approach could help to understand or help to predict more accurately the impact of environmental factors triggering outbreaks and to design either new prophylactic or therapeutic strategic based on microorganism or microbial metabolites. In a more holistic approach, this also could help to understand if changes in the environmental microbiota are sensed by the fish or other species which should help to properly assess their impact on the aquaculture ecosystem.

4.2 Prediction systems: an ecosystem approach

FAO's ecosystem approach to aquaculture [192] is a "strategy for the integration of the activity within the wider ecosystem such that it promotes sustainable development, equity, and resilience of interlinked social-ecological systems." The so-called aquaculture ecosystem of southern Chile is shared by multiple users, notably by the salmon and mussel industry and a significant part of the national fishermen task force that is concentrated in Region X (De Los Lagos Region or Lake District). The latter depends on seafood collection and commercialization for their subsistence, and so any serious ecosystem perturbation ends up in conflicting situations affecting all users, including tourism, also an important player depending on the marine ecosystem. One of the most serious harmful algal blooms (HABs) of *Pseudochattonella verruculosa* occurred in the austral summer of 2016 (February–April) killing nearly 12% of the Chilean salmon production (106,000 tons), causing severe mortality of other fish and shellfish in the coastal waters and interior sea of western Patagonia [193]. This event exemplifies the inherent complexity of ecosystem perturbations and its socioecological consequences. But not only users like aquaculture producers should be blamed by such perturbations since climatic change seems to have created the oceanographic conditions that amplified this HAB event. Indeed, the event was associated to El Niño and the climatic and oceanographic conditions associated to it [193]. In spite of the ongoing monitoring protocols carried out by different institutions, it has been difficult to understand what factors influence the diversity and abundance of harmful microalgae population, which is understandable due to the dynamic and complexity of the marine ecosystem, so a HAB event cannot be understood from the analysis of few variables. Metagenomic studies offer new insights into the complexity of the marine ecosystem and HAB events by allowing a deeper view to the microbial diversity that cannot be approached by the traditional microscopic analysis often used for microalgae identification. Additionally, different sorts of interactions can be discovered at all levels, particularly between microbes and microalgae [194–198]. It is now known that some bacterial populations could promote the growth of specific harmful microalgae in species, while some bacteria related to disease in fish or mollusks could also promote blooms. But, also virus controls the abundance and activity of microbial populations and microalgae in nature. In short, the ecosystem associated to HABs is complex and needs a more holistic or integrated approach. The term holobiome has been suggested to address this complexity: holo = entity; biome = biological community. An ongoing project funded by the Japanese government in Chile integrates different Japanese and Chilean universities, the Instituto de Fomento Pesquero (IFOP), and private and governmental bodies, under the holobiome concept with the goal to shed light on the mechanisms involved on algal bloom formation and, at the same time, predict HAB blooms (www.mach-satreps.org/en/).

5. Conclusions

1. Chilean aquaculture is the focus of this review considering its local and world relevance, particularly the farming of exotic salmonids, one of the highly regarded productive clusters.

2. Given the intensive farming strategy, the cost-effectiveness of the salmon industry is affected by three critical situations: the outbreak of mortal viruses, the occurrence and persistence of bacterial diseases, and the scale of antibiotic usage, one the highest in the world, with serious environmental consequences.

3. Understanding the interaction between host (fish), microbiota, and environmental microorganisms could be a key factor to develop rational strategies to improve the productivity by increasing the resistance to infection, reducing the use of antibiotics and their negative environmental impact.

4. The application of metagenomics to understand host (fish)-microbiota-environment interaction in salmon farming is limited, a reality opposed to the economic gains of the industry. Nevertheless, it has contributed to evaluate (i) the skin (the first barrier to pathogens or parasites) and gut microbiota of farmed salmonid subject to different nutritional and growth conditions, (ii) the function of the microbiota as antagonist against pathogens, and (iii) the potential function in the nutrition and environmental adaption.

5. Metagenomics has contributed to the identification of probiotic bacteria in both salmonids and marine fish species with significant benefits to the industry.

6. Metagenomics is essential to assess and predict critical environmental perturbations affecting fish and mollusk species like harmful algal blooms (HABs). Addressing such a problem requires to have a holistic view of the ecosystem components and their interactions, including the microbial diversity.

7. The sustainable expansion and diversification of Chilean aquaculture require to incorporate metagenomics and other omics tools to successfully deal with current and new diseases. This is required for aquaculture to be a really efficient and environment-friendly industry.

Acknowledgements

AG and GG acknowledge support from to the project GIAP-02 "Producción Acuícola Sustentable" and MACH (SATREPS-JICA). MT acknowledges support from Project Mecesup USA1799 and 021971GM_DAS. The section on prediction systems is inspired by the ongoing SATREP-JICA project (Japan-Chile cooperation): "Development of harmful algal bloom monitoring methods and forecast system for sustainable aquaculture and coastal fisheries in Chile" (MACH, GG Osorno responsible).

Conflict of interest

The authors declared no conflict of interest.

Notes/thanks/other declarations

We would like to thank Susan Smalley for the correction of this chapter.

Author details

Mario Tello[1*], Natalia Valdes[1], Rodrigo Vargas[3], Joselin Rojas[1], Mick Parra[1], Gonzalo Gajardo[2,4] and Alex Gonzalez[3,4]

1 Centro de Biotecnología Acuícola, Universidad de Santiago de Chile, Alameda, Santiago, Chile

2 Departamento de Ciencias Biológicas y Biodiversidad, Laboratorio de Genética, Acuicultura and Biodiversidad, Universidad de los Lagos, Osorno, Chile

3 Departamento de Ciencias Biológicas y Biodiversidad, Laboratorio de Microbiología Ambiental y Extremofilos, Universidad de los Lagos, Osorno, Chile

4 Grupo de Investigación en Producción Acuícola Sustentable, Universidad de los Lagos, Osorno, Chile

*Address all correspondence to: mario.tello@usach.cl

References

[1] Sorgeloos P. AQUACULTURE: The blue biotechnology of the future. World Aquaculture Society. 2013;**35**:18-27

[2] FAO. SOFIA 2018—State of Fisheries and Aquaculture in the World 2018. 2018. Available from: http://www.fao.org/state-of-fisheries-aquaculture

[3] Duarte CM, Marba N, Holmer M. ECOLOGY: Rapid domestication of marine species. Science. 2007;**316**(5823):382-383. Available from: http://www.ncbi.nlm.nih.gov/pubmed/17446380

[4] Soumya Haldar SC. Vibrio related diseases in aquaculture and development of rapid and accurate identification methods. Journal of Marine Science: Research & Development. 2012;**s1**(1):1-7. Available from: https://www.omicsonline.org/vibrio-related-diseases-in-aquaculture-and-development-of-rapid-and-accurate-identification-methods-2155-9910.S1-002.php?aid=6125

[5] Defoirdt T, Halet D, Vervaeren H, Boon N, Van de Wiele T, Sorgeloos P, et al. The bacterial storage compound poly-?-hydroxybutyrate protects Artemia franciscana from pathogenic Vibrio campbellii. Environmental Microbiology. 2007;**9**(2):445-452. Available from: http://www.ncbi.nlm.nih.gov/pubmed/17222142

[6] Baruah K, Huy TT, Norouzitallab P, Niu Y, Gupta SK, De Schryver P, et al. Probing the protective mechanism of poly-ß-hydroxybutyrate against vibriosis by using gnotobiotic Artemia franciscana and Vibrio campbellii as host-pathogen model. Scientific Reports. 2015;**5**:9427. Available from: http://www.ncbi.nlm.nih.gov/pubmed/25822312

[7] Miranda CD, Godoy FA, Lee MR. Current status of the use of antibiotics and the antimicrobial resistance in the Chilean Salmon farms. Frontiers in Microbiology. 2018;**9**:1284. Available from: http://www.ncbi.nlm.nih.gov/pubmed/29967597

[8] SERNAPESCA. Estadísticas | Servicio Nacional de Pesca y Acuicultura. 2017. Available from: http://www.sernapesca.cl/informes/estadisticas

[9] García de Leaniz C, Gajardo G, Consuegra S. From Best to Pest: Changing perspectives on the impact of exotic salmonids in the southern hemisphere. Systematics and Biodiversity. 2010;**8**(4):447-459. Available from: http://www.tandfonline.com/doi/abs/10.1080/14772000.2010.537706

[10] Gajardo G, Laikre L. Chilean aquaculture boom is based on exotic Salmon resources: A conservation paradox. Conservation Biology. 2003;**17**(4):1173-1174. Available from: http://doi.wiley.com/10.1046/j.1523-1739.2003.02351.x

[11] Martínez-Porchas M, Vargas-Albores F. Microbial metagenomics in aquaculture: A potential tool for a deeper insight into the activity. Reviews in Aquaculture. 2017;**9**(1):42-56. Available from: http://doi.wiley.com/10.1111/raq.12102

[12] FAO. Fisheries and Aquaculture Department. Geographic information—Introduction. In: FAO Fisheries and Aquaculture Department. 2012. Available from: http://www.fao.org/fi/common/format/popUpCitation.jsp?type=citation

[13] Escobar-Dodero J, Kinsley A, Perez AM, Ibarra R, Tello A, Monti G, et al. Risk factors for infectious pancreatic necrosis in farmed Chilean Atlantic salmon (*Salmo salar* L.) from 2010 to 2013;**167**:182-189. Preventive Veterinary Medicine. 2018. Available from: http://www.ncbi.nlm.nih.gov/pubmed/29891102

[14] Alvial A, Kibenge F, Forster CJ, Burgos JM. The Recovery of the Chilean Salmon Industry the ISA Crisis and Its Consequences and Lessons. 2012. Available from: https://www.aquaculturealliance.org/wp-content/uploads/2015/02/GAA_ISA-Report.pdf

[15] SERNAPESCA. Informe Sanitario de Salmonicultura en Centros Marinos 2017. 2018. Available from: www.sernapesca.cl

[16] Duchaud E, Rochat T, Habib C, Barbier P, Loux V, Guérin C, et al. Genomic diversity and evolution of the fish pathogen Flavobacterium psychrophilum. Frontiers in Microbiology. 2018;**9**:138. Available from: http://www.ncbi.nlm.nih.gov/pubmed/29467746

[17] Bayliss SC, Verner-Jeffreys DW, Ryder D, Suarez R, Ramirez R, Romero J, et al. Genomic epidemiology of the commercially important pathogen Renibacterium salmoninarum within the Chilean salmon industry. Microbial Genomics. 2018;**4**(9):e000201. Available from: http://www.ncbi.nlm.nih.gov/pubmed/30040063

[18] Houston RD, Gheyas A, Hamilton A, Guy DR, Tinch AE, Taggart JB, et al. Detection and confirmation of a major QTL affecting resistance to infectious pancreatic necrosis (IPN) in Atlantic Salmon (*Salmo Salar*). In: Animal Genomics for Animal Health. Basel: KARGER; 2008. pp. 199-204. Available from: http://www.ncbi.nlm.nih.gov/pubmed/18817302

[19] Munang'andu HM, Fredriksen BN, Mutoloki S, Brudeseth B, Kuo T-Y, Marjara IS, et al. Comparison of vaccine efficacy for different antigen delivery systems for infectious pancreatic necrosis virus vaccines in Atlantic salmon (Salmo salar L.) in a cohabitation challenge model. Vaccine. 2012;**30**(27):4007-4016. Available from: http://www.ncbi.nlm.nih.gov/pubmed/22537985

[20] Cottet L, Cortez-San Martin M, Tello M, Olivares E, Rivas-Aravena A, Vallejos E, et al. Bioinformatic analysis of the genome of infectious salmon anemia virus associated with outbreaks with high mortality in Chile. Journal of Virology. 2010;**84**(22):11916-11928. Available from: http://www.pubmedcentral.nih.gov/articlerender.fcgi?artid=2977892&tool=pmcentrez&rendertype=abstract

[21] Cartagena J, Tambley C, Sandino AM, Spencer E, Tello M. Detection of piscine orthoreovirus in farmed rainbow trout from Chile. Aquaculture. 2018;**493**:79-84

[22] SERNAPESCA. Estudio Sobre Detección de Piscine Reovirus en Centros de Cultivos de Salmónidos Ubicados en Mar en la X, XI y XII región. Santiago; 2012

[23] Meixner D. Determinación De La Dosis Letal 50 (Dl50) De Una Cepa Nacional De Flavobacterium Psychrophilum En Trucha Arcoiris (*Oncorhynchus Mykiss*). UACH. 2007. Available from: http://cybertesis.uach.cl/tesis/uach/2007/fvm515d/doc/fvm515d.pdf

[24] Martinez A. Enfermedades infecciosas—SalmonExpert.cl. Salmon expert. 2018. Available from: https://www.salmonexpert.cl/article/enfermedades-infecciosas/

[25] SERNAPESCA. Informe Sobre Uso de Antimicrobianos en la Salmonicultura Nacional 1° Semestre 2018. 2018. Available from: http://www.sernapesca.cl/sites/default/files/informe_sobre_uso_de_antimicrobianos_en_la_salmonicultura_nacional_primer_semestre_2018.pdf

[26] Lozano I, Díaz NF, Muñoz S, Riquelme C. Antibiotics in Chilean

aquaculture: A review. In: Antibiotic Use in Animals. Rijeka: InTech; 2018. Available from: http://www.intechopen.com/books/antibiotic-use-in-animals/antibiotics-in-chilean-aquaculture-a-review

[27] de la Fuente M, Miranda C, Faúndez V. Bacteriología asociada al cultivo de moluscos en Chile: Avances y perspectivas. Revista de Biología Marina y Oceanografía. 2015;**50**(1):01-12. Available from: http://www.scielo.cl/scielo.php?script=sci_arttext&pid=S0718-19572015000100001&lng=en&nrm=iso&tlng=en

[28] Lohrmann KB, Bustos E, Rojas R, Navarrete F, Robotham H, Bignell J. Histopathological assessment of the health status of Mytilus chilensis (Hupé 1854) in southern Chile. Aquaculture. 2019;**503**:40-50. Available from: https://www.sciencedirect.com/science/article/abs/pii/S0044848618312092

[29] Methé BA, Nelson KE, Pop M, Creasy HH, Giglio MG, Huttenhower C, et al. A framework for human microbiome research. Nature. 2012;**486**(7402):215-221. Available from: http://www.nature.com/doifinder/10.1038/nature11209

[30] Huttenhower C, Gevers D, Knight R, Abubucker S, Badger JH, Chinwalla AT, et al. Structure, function and diversity of the healthy human microbiome. Nature. 2012;**486**(7402):207-214. Available from: http://www.ncbi.nlm.nih.gov/pubmed/22699609

[31] Sender R, Fuchs S, Milo R, Berg RD, Bianconi E, Piovesan A, et al. Are we really vastly outnumbered? Revisiting the ratio of bacterial to host cells in humans. Cell. 2016;**164**(3): 337-340. Available from: http://linkinghub.elsevier.com/retrieve/pii/S0092867416000532

[32] Staley JT, Konopka A. Measurement of in situ activities of nonphotosynthetic microorganisms in aquatic and terrestrial habitats. Annual Review of Microbiology. 1985;**39**:321-346. Available from: http://www.ncbi.nlm.nih.gov/pubmed/3904603

[33] Schmidt TM, DeLong EF, Pace NR. Analysis of a marine picoplankton community by 16S rRNA gene cloning and sequencing. Journal of Bacteriology. 1991;**173**(14):4371-4378. Available from: http://jb.asm.org/content/173/14/4371?ijkey=96a109db860b9fff3e45b056e2f3b48d3458489c&keytype2=tf_ipsecsha

[34] Muyzer G. DGGE/TGGE a method for identifying genes from natural ecosystems. Current Opinion in Microbiology. 1999;**2**(3):317-322. Available from: http://www.ncbi.nlm.nih.gov/pubmed/10383868

[35] Nikolaki S, Tsiamis G. Microbial diversity in the era of omic technologies. BioMed Research International. 2013;**2013**:958719. Available from: http://www.pubmedcentral.nih.gov/articlerender.fcgi?artid=3821902&tool=pmcentrez&rendertype=abstract

[36] Moeller AH, Caro-Quintero A, Mjungu D, Georgiev AV, Lonsdorf EV, Muller MN, et al. Cospeciation of gut microbiota with hominids. Science. 2016;**353**(6297):380-382. Available from: http://www.ncbi.nlm.nih.gov/pubmed/27463672

[37] Blaser MJ, Webb GF. Host demise as a beneficial function of indigenous microbiota in human hosts. MBio. 2014;**5**(6):e02262-e02214. Available from: http://mbio.asm.org/lookup/doi/10.1128/mBio.02262-14

[38] Brito IL, Yilmaz S, Huang K, Xu L, Jupiter SD, Jenkins AP, et al. Mobile genes in the human microbiome are structured from global to individual scales. Nature. 2016;**535**(7612):435-439.

Available from: http://www.nature.com/doifinder/10.1038/nature18927

[39] Clarke G, Stilling RM, Kennedy PJ, Stanton C, Cryan JF, Dinan TG. Mini review: Gut microbiota: The neglected endocrine organ. Molecular Endocrinology. 2014;**28**(8):1221-1238. Available from: http://www.ncbi.nlm.nih.gov/pubmed/24892638

[40] Salvucci E. Microbiome, holobiont and the net of life. Critical Reviews in Microbiology. 2016;**42**(3):485-494. Available from: http://www.ncbi.nlm.nih.gov/pubmed/25430522

[41] Hoffmann AR, Proctor LM, Surette MG, Suchodolski JS. The microbiome: The trillions of microorganisms that maintain health and cause disease in humans and companion animals. Veterinary Pathology. 2015;**53**(1):10-21. Available from: http://www.ncbi.nlm.nih.gov/pubmed/26220947

[42] Chiu L, Bazin T, Truchetet M-E, Schaeverbeke T, Delhaes L, Pradeu T. Protective microbiota: From localized to long-reaching Co-immunity. Frontiers in Immunology. 2017;**8**:1678. Available from: http://www.ncbi.nlm.nih.gov/pubmed/29270167

[43] Cawthon CR, de La Serre CB. Gut bacteria interaction with vagal afferents. Brain Research. 2018;**1693**(Pt B):134-139. Available from: http://www.ncbi.nlm.nih.gov/pubmed/29360469

[44] Turnbaugh PJ, Ley RE, Mahowald MA, Magrini V, Mardis ER, Gordon JI. An obesity-associated gut microbiome with increased capacity for energy harvest. Nature. 2006;**444**(7122):1027-1031. DOI: 10.1038/nature05414

[45] Tomkovich S, Jobin C. Microbiota and host immune responses: A love-hate relationship. Immunology. 2016;**147**(1):1-10. Available from: http://www.ncbi.nlm.nih.gov/pubmed/26439191

[46] Gensollen T, Iyer SS, Kasper DL, Blumberg RS. How colonization by microbiota in early life shapes the immune system. Science. 2016;**352**(6285):539-544. Available from: http://www.ncbi.nlm.nih.gov/pubmed/27126036

[47] Llewellyn MS, Boutin S, Hoseinifar SH, Derome N. Teleost microbiomes: The state of the art in their characterization, manipulation and importance in aquaculture and fisheries. Frontiers in Microbiology. 2014;**5**:207. Available from: http://www.pubmedcentral.nih.gov/articlerender.fcgi?artid=4040438&tool=pmcentrez&rendertype=abstract

[48] Xiong J-B, Nie L, Chen J. Current understanding on the roles of gut microbiota in fish disease and immunity. Zoological Research. 2018;**40**(2):70-76. Available from: http://www.ncbi.nlm.nih.gov/pubmed/29976843

[49] Zarkasi KZ, Taylor RS, Abell GCJ, Tamplin ML, Glencross BD, Bowman JP. Atlantic Salmon (*Salmo salar* L.) gastrointestinal microbial community dynamics in relation to Digesta properties and diet. Microbial Ecology. 2016;**71**(3):589-603. Available from: http://link.springer.com/10.1007/s00248-015-0728-y

[50] Schmidt V, Amaral-Zettler L, Davidson J, Summerfelt S, Good C. Influence of fishmeal-free diets on microbial communities in Atlantic Salmon (Salmo salar) recirculation aquaculture systems. Applied and Environmental Microbiology. 2016;**82**(15):4470-4481. Available from: http://www.ncbi.nlm.nih.gov/pubmed/27129964

[51] Zarkasi KZ, Taylor RS, Glencross BD, Abell GCJ, Tamplin ML, Bowman JP. In vitro characteristics of an Atlantic

salmon (*Salmo salar* L.) hind gut microbial community in relation to different dietary treatments. Research in Microbiology. 2017;**168**(8):751-759. Available from: http://linkinghub.elsevier.com/retrieve/pii/S0923250817301304

[52] Gajardo K, Jaramillo-Torres A, Kortner TM, Merrifield DL, Tinsley J, Bakke AM, et al. Alternative protein sources in the diet modulate microbiota and functionality in the distal intestine of Atlantic Salmon (*Salmo salar*). Björkroth J, editor. Applied and Environmental Microbiology. 2017;**83**(5):e02615-e02616. Available from: http://aem.asm.org/lookup/doi/10.1128/AEM.02615-16

[53] Catalán N, Villasante A, Wacyk J, Ramírez C, Romero J. Fermented soybean meal increases lactic acid bacteria in gut microbiota of Atlantic Salmon (*Salmo salar*). Probiotics and Antimicrobial Proteins. 2017;**10**(3):566-576. Available from: http://www.ncbi.nlm.nih.gov/pubmed/29274013

[54] Rimoldi S, Terova G, Ascione C, Giannico R, Brambilla F. Next generation sequencing for gut microbiome characterization in rainbow trout (*Oncorhynchus mykiss*) fed animal by-product meals as an alternative to fishmeal protein sources. Soengas JL, editor. PLoS One 2018;**13**(3):e0193652. Available from: http://dx.plos.org/10.1371/journal.pone.0193652

[55] Mente E, Nikouli E, Antonopoulou E, Martin SAM, Kormas KA. Core versus diet-associated and postprandial bacterial communities of the rainbow trout (*Oncorhynchus mykiss*) midgut and faeces. Biology Open. 2018;**7**(6):bio034397. Available from: http://bio.biologists.org/lookup/doi/10.1242/bio.034397

[56] Michl SC, Ratten J-M, Beyer M, Hasler M, LaRoche J, Schulz C. The malleable gut microbiome of juvenile rainbow trout (*Oncorhynchus mykiss*): Diet-dependent shifts of bacterial community structures. Prunet P, editor. PLoS One. 2017;**12**(5):e0177735. Available from: http://dx.plos.org/10.1371/journal.pone.0177735

[57] Hatje E, Neuman C, Stevenson H, Bowman JP, Katouli M. Population dynamics of vibrio and pseudomonas species isolated from farmed Tasmanian Atlantic salmon (*Salmo salar* L.): A seasonal study. Microbial Ecology. 2014;**68**(4):679-687. Available from: http://www.ncbi.nlm.nih.gov/pubmed/25027277

[58] Zarkasi KZ, Abell GCJ, Taylor RS, Neuman C, Hatje E, Tamplin ML, et al. Pyrosequencing-based characterization of gastrointestinal bacteria of Atlantic salmon (*Salmo salar* L.) within a commercial mariculture system. Journal of Applied Microbiology. 2014;**117**(1):18-27. Available from: http://www.ncbi.nlm.nih.gov/pubmed/24698479

[59] Llewellyn MS, McGinnity P, Dionne M, Letourneau J, Thonier F, Carvalho GR, et al. The biogeography of the Atlantic salmon (*Salmo salar*) gut microbiome. The ISME Journal. 2016;**10**(5):1280-1284. Available from: http://www.ncbi.nlm.nih.gov/pubmed/26517698

[60] Dehler CE, Secombes CJ, Martin SAM. Environmental and physiological factors shape the gut microbiota of Atlantic salmon parr (*Salmo salar* L.). Aquaculture. 2017;**467**:149-157. Available from: http://www.ncbi.nlm.nih.gov/pubmed/28111483

[61] He X, Chaganti SR, Heath DD. Population-specific responses to interspecific competition in the gut microbiota of two Atlantic Salmon (Salmo salar) populations. Microbial Ecology. 2018;**75**(1):140-151. Available

from: http://www.ncbi.nlm.nih.gov/pubmed/28714057

[62] Dehler CE, Secombes CJ, Martin SAM. Seawater transfer alters the intestinal microbiota profiles of Atlantic salmon (Salmo salar L.). Scientific Reports. 2017;7(1):13877. Available from: http://www.nature.com/articles/s41598-017-13249-8

[63] Lyons PP, Turnbull JF, Dawson KA, Crumlish M. Phylogenetic and functional characterization of the distal intestinal microbiome of rainbow trout Oncorhynchus mykiss from both farm and aquarium settings. Journal of Applied Microbiology. 2017;**122**(2):347-363. Available from: http://doi.wiley.com/10.1111/jam.13347

[64] Uren Webster TM, Consuegra S, Hitchings M, Garcia de Leaniz C. Inter-population variation in the Atlantic salmon microbiome reflects environmental and genetic diversity. Applied and Environmental Microbiology. 2018;**84**(16):e00691-18. Available from: http://aem.asm.org/lookup/doi/10.1128/AEM.00691-18

[65] Lokesh J, Kiron V, Marchesi JR, Ravel J, Gomez D, Sunyer JO, et al. Transition from freshwater to seawater reshapes the skin-associated microbiota of Atlantic salmon. Scientific Reports. 2016;**6**:19707. Available from: http://www.nature.com/articles/srep19707

[66] Lokesh J, Kiron V, Sipkema D, Fernandes JMO, Moum T. Succession of embryonic and the intestinal bacterial communities of Atlantic salmon (*Salmo salar*) reveals stage-specific microbial signatures. Microbiology. 2018;**8**(4):e00672. Available from: http://www.ncbi.nlm.nih.gov/pubmed/29897674

[67] Rudi K, Angell IL, Pope PB, Vik JO, Sandve SR, Snipen L-G. Stable Core gut microbiota across the freshwater-to-saltwater transition for farmed Atlantic Salmon. Drake HL, editor. Applied and Environmental Microbiology. 2018;**84**(2):e01974-e01917. Available from: http://aem.asm.org/lookup/doi/10.1128/AEM.01974-17

[68] Gajardo K, Rodiles A, Kortner TM, Krogdahl Å, Bakke AM, Merrifield DL, et al. A high-resolution map of the gut microbiota in Atlantic salmon (*Salmo salar*): A basis for comparative gut microbial research. Scientific Reports. 2016;**6**:30893. Available from: http://www.nature.com/articles/srep30893

[69] Lowrey L, Woodhams DC, Tacchi L, Salinas I. Topographical mapping of the rainbow trout (*Oncorhynchus mykiss*) microbiome reveals a diverse bacterial community in the skin with antifungal properties. Applied and Environmental Microbiology. 2015;**81**(19):6915-6925. Available from: http://www.ncbi.nlm.nih.gov/pubmed/26209676

[70] Navarrete P, Mardones P, Opazo R, Espejo R, Romero J. Oxytetracycline treatment reduces bacterial diversity of intestinal microbiota of Atlantic salmon. Journal of Aquatic Animal Health. 2008;**20**(3):177-183. Available from: http://www.ncbi.nlm.nih.gov/pubmed/18942594

[71] Kamada N, Kim Y-G, Sham HP, Vallance BA, Puente JL, Martens EC, et al. Regulated virulence controls the ability of a pathogen to compete with the gut microbiota. Science. 2012;**336**(6086):1325-1329. Available from: http://www.pubmedcentral.nih.gov/articlerender.fcgi?artid=3439148&tool=pmcentrez&rendertype=abstract

[72] Donia MS, Fischbach MA. Small molecules from the human microbiota. Science. 2015;**349**(6246):1254766-1254766. Available from: http://www.ncbi.nlm.nih.gov/pubmed/26206939

[73] Liu Y, de Bruijn I, Jack ALH, Drynan K, van den Berg AH, Thoen E, et al. Deciphering microbial landscapes of

fish eggs to mitigate emerging diseases. The ISME Journal. 2014;**8**(10):2002-2014. Available from: http://www.ncbi.nlm.nih.gov/pubmed/24671087

[74] Banerjee G, Ray AK. The advancement of probiotics research and its application in fish farming industries. Research in Veterinary Science. 2017;**115**:66-77. Available from: http://www.ncbi.nlm.nih.gov/pubmed/28157611

[75] Zorriehzahra MJ, Delshad ST, Adel M, Tiwari R, Karthik K, Dhama K, et al. Probiotics as beneficial microbes in aquaculture: An update on their multiple modes of action: A review. The Veterinary Quarterly. 2016;**36**(4): 228-241. Available from: http://www.ncbi.nlm.nih.gov/pubmed/27075688

[76] Boutin S, Bernatchez L, Audet C, Derôme N. Network analysis highlights complex interactions between pathogen, host and commensal microbiota. PLoS One. 2013;**8**(12):e84772. Available from: http://www.pubmedcentral.nih.gov/articlerender.fcgi?artid=3871659&tool=pmcentrez&rendertype=abstract

[77] Einar R, Zhigang Z, Suxu H, Rolf EO. Effect of stress on intestinal microbiota of Arctic charr, Atlantic salmon, rainbow trout and Atlantic cod: A review. African Journal of Microbiology Research. 2014;**8**(7):609-618. Available from: http://www.academicjournals.org/journal/AJMR/article-abstract/D84210A43012

[78] Borrelli L, Aceto S, Agnisola C, De Paolo S, Dipineto L, Stilling RM, et al. Probiotic modulation of the microbiota-gut-brain axis and behaviour in zebrafish. Scientific Reports. 2016;**6**:30046. Available from: http://www.ncbi.nlm.nih.gov/pubmed/27416816

[79] Tapia-Paniagua ST, Vidal S, Lobo C, Prieto-Álamo MJ, Jurado J, Cordero H, et al. The treatment with the probiotic Shewanella putrefaciens Pdp11 of specimens of Solea senegalensis exposed to high stocking densities to enhance their resistance to disease. Fish & Shellfish Immunology. 2014;**41**(2):209-221. Available from: http://www.ncbi.nlm.nih.gov/pubmed/25149590

[80] Taoka Y, Maeda H, Jo J-Y, Jeon M-J, Bai SC, Lee W-J, et al. Growth, stress tolerance and non-specific immune response of Japanese flounder Paralichthys olivaceus to probiotics in a closed recirculating system. Fisheries Science. 2006;**72**(2):310-321. Available from: http://link.springer.com/10.1111/j.1444-2906.2006.01152.x

[81] Merrifield DL, Dimitroglou A, Foey A, Davies SJ, Baker RTM, Bøgwald J, et al. The current status and future focus of probiotic and prebiotic applications for salmonids. Aquaculture. 2010;**302**:(1-2):1-18. Available from: https://www.sciencedirect.com/science/article/abs/pii/S0044848610000955

[82] Rawls JF, Samuel BS, Gordon JI. Gnotobiotic zebrafish reveal evolutionarily conserved responses to the gut microbiota. Proceedings of the National Academy of Sciences of the United States of America. 2004;**101**(13):4596-4601. Available from: http://www.pubmedcentral.nih.gov/articlerender.fcgi?artid=384792&tool=pmcentrez&rendertype=abstract

[83] Semova I, Carten JD, Stombaugh J, Mackey LC, Knight R, Farber SA, et al. Microbiota regulate intestinal absorption and metabolism of fatty acids in the zebrafish. Cell Host & Microbe. 2012;**12**(3):277-288. Available from: http://www.ncbi.nlm.nih.gov/pubmed/22980325

[84] Bates JM, Mittge E, Kuhlman J, Baden KN, Cheesman SE, Guillemin K. Distinct signals from the microbiota promote different aspects of zebrafish gut differentiation. Developmental Biology. 2006;**297**(2):374-386. Available

from: http://www.ncbi.nlm.nih.gov/pubmed/16781702

[85] Cheesman SE, Neal JT, Mittge E, Seredick BM, Guillemin K. Epithelial cell proliferation in the developing zebrafish intestine is regulated by the Wnt pathway and microbial signaling via Myd88. Proceedings of the National Academy of Sciences. 2011;**108**(Supplement_1):4570-4577. Available from: http://www.ncbi.nlm.nih.gov/pubmed/20921418

[86] Stephens WZ, Burns AR, Stagaman K, Wong S, Rawls JF, Guillemin K, et al. The composition of the zebrafish intestinal microbial community varies across development. The ISME Journal. 2016;**10**(3):644-654. Available from: http://www.ncbi.nlm.nih.gov/pubmed/26339860

[87] Barros-Becker F, Romero J, Pulgar A, Feijóo CG. Persistent oxytetracycline exposure induces an inflammatory process that improves regenerative capacity in zebrafish larvae. PLoS One. 2012;7(5):e36827. Available from: http://www.pubmedcentral.nih.gov/articlerender.fcgi?artid=3349639&tool=pmcentrez&rendertype=abstract

[88] Galindo-Villegas J, Garcia-Moreno D, de Oliveira S, Meseguer J, Mulero V. Regulation of immunity and disease resistance by commensal microbes and chromatin modifications during zebrafish development. Proceedings of the National Academy of Sciences. 2012;**109**(39):E2605-E2614. Available from: http://www.ncbi.nlm.nih.gov/pubmed/22949679

[89] Kamada N, Chen GY, Inohara N, Núñez G. Control of pathogens and pathobionts by the gut microbiota. Nature Immunology. 2013;**14**(7):685-690. Available from: http://www.ncbi.nlm.nih.gov/pubmed/23778796

[90] Stagaman K, Burns AR, Guillemin K, Bohannan BJ. The role of adaptive immunity as an ecological filter on the gut microbiota in zebrafish. The ISME Journal. 2017;**11**(7):1630-1639. Available from: http://www.ncbi.nlm.nih.gov/pubmed/28304369

[91] Hai N. The use of probiotics in aquaculture. Journal of Applied Microbiology. 2015;**119**(4):917-935. Available from: http://www.ncbi.nlm.nih.gov/pubmed/26119489

[92] Reveco FE, Øverland M, Romarheim OH, Mydland LT. Intestinal bacterial community structure differs between healthy and inflamed intestines in Atlantic salmon (Salmo salar L.). Aquaculture. 2014;**420**:262-269

[93] Landeira-Dabarca A, Sieiro C, Álvarez M. Change in food ingestion induces rapid shifts in the diversity of microbiota associated with cutaneous mucus of Atlantic salmon Salmo salar. Journal of Fish Biology. 2013;**82**(3):893-906. Available from: http://www.ncbi.nlm.nih.gov/pubmed/23464550

[94] Mu Y, Ding F, Cui P, Ao J, Hu S, Chen X. Transcriptome and expression profiling analysis revealed changes of multiple signaling pathways involved in immunity in the large yellow croaker during Aeromonas hydrophila infection. BMC Genomics. 2010;**11**:506. Available from: http://www.pubmedcentral.nih.gov/articlerender.fcgi?artid=2997002&tool=pmcentrez&rendertype=abstract

[95] Lazado CC, Caipang CMA. Mucosal immunity and probiotics in fish. Fish & Shellfish Immunology. 2014;**39**(1):78-89. Available from: http://www.ncbi.nlm.nih.gov/pubmed/24795079

[96] Miranda CD. Antimicrobial resistance associated with salmonid farming. In: Antimicrobial Resistance in the Environment. Hoboken, NJ, USA: John Wiley & Sons, Inc.; 2011. pp. 423-451. Available from: http://doi.wiley.com/10.1002/9781118156247.ch22

[97] Du L, Liu W. Occurrence, fate, and ecotoxicity of antibiotics in agro-ecosystems. A review. Agronomy for Sustainable Development. 2012;**32**(2):309-327. Available from: http://link.springer.com/10.1007/s13593-011-0062-9

[98] Kemper N. Veterinary antibiotics in the aquatic and terrestrial environment. Ecological Indicators. 2008;**8**(1):1-13. Available from: https://www.sciencedirect.com/science/article/pii/S1470160X07000647

[99] Hollis A, Ahmed Z. The path of least resistance: Paying for antibiotics in non-human uses. Health Policy (New York). 2014;**118**(2):264-270. Available from: http://www.ncbi.nlm.nih.gov/pubmed/25240271

[100] SERNAPESCA. Informe Sobre Uso De Antimicrobianos En La Salmonicultura Nacional Año. Vol. 2017. 2017. Available from: http://www.sernapesca.cl/sites/default/files/informe_sobre_uso_de_antimicrobianos_2017_0.pdf

[101] Rozas M, Enríquez R. Piscirickettsiosis and Piscirickettsia salmonis in fish: A review. Journal of Fish Diseases. 2014;**37**(3):163-188. Available from: http://www.ncbi.nlm.nih.gov/pubmed/24279295

[102] Romero J, Gloria C, Navarrete P. Antibiotics in aquaculture – Use, abuse and alternatives. In: Health and Environment in Aquaculture. Rijeka: InTech; 2012. Available from: http://www.intechopen.com/books/health-and-environment-in-aquaculture/antibiotics-in-aquaculture-use-abuse-and-alternatives

[103] Camussetti M, Gallardo A, Aguilar D, Larenas J. Análisis de los costos por la utilización de quimioterápicos y vacunas en la salmonicultura. SalmonExpert.cl. Salmonexpert. 2015;**4**:46-49. Available from: https://www.salmonexpert.cl/article/an-aacute-lisis-de-los-costos-por-la-utilizaci-oacute-n-de-quimioter-aacute-picos-y-vacunas-en-la-salmonicultura

[104] Wilson AB. MHC and adaptive immunity in teleost fishes. Immunogenetics. 2017;**69**(8-9):521-528. Available from: http://www.ncbi.nlm.nih.gov/pubmed/28695284

[105] Fortt ZA, Cabello CF, Buschmann RA. Residuos de tetraciclina y quinolonas en peces silvestres en una zona costera donde se desarrolla la acuicultura del salmón en Chile. Revista Chilena de Infectologia. 2007;**24**(1):14-18. Available from: http://www.scielo.cl/scielo.php?script=sci_arttext&pid=S0716-10182007000100002&lng=en&nrm=iso&tlng=en

[106] Yang J, Wang C, Shu C, Liu L, Geng J, Hu S, et al. Marine sediment bacteria harbor antibiotic resistance genes highly similar to those found in human pathogens. Microbial Ecology. 2013;**65**(4):975-981. Available from: http://www.ncbi.nlm.nih.gov/pubmed/23370726

[107] Higuera-Llantén S, Vásquez-Ponce F, Barrientos-Espinoza B, Mardones FO, Marshall SH, Olivares-Pacheco J. Extended antibiotic treatment in salmon farms select multiresistant gut bacteria with a high prevalence of antibiotic resistance genes. Luo Y, editor. PLoS One. 2018;**13**(9):e0203641. Available from: https://dx.plos.org/10.1371/journal.pone.0203641

[108] Dupraz V, Stachowski-Haberkorn S, Ménard D, Limon G, Akcha F, Budzinski H, et al. Combined effects of antifouling biocides on the growth of three marine microalgal species. Chemosphere. 2018;**209**:801-814. Available from: https://www.sciencedirect.com/science/article/pii/S0045653518312141

[109] Claus SP, Guillou H, Ellero-Simatos S. The gut microbiota: A major player in the toxicity of environmental pollutants? npj Biofilms and Microbiomes. 2016;**2**:16003. Available from: http://www.ncbi.nlm.nih.gov/pubmed/28721242

[110] Meng X-L, Li S, Qin C-B, Zhu Z-X, Hu W-P, Yang L-P, et al. Intestinal microbiota and lipid metabolism responses in the common carp (*Cyprinus carpio* L.) following copper exposure. Ecotoxicology and Environmental Safety. 2018;**160**:257-264. Available from: http://www.ncbi.nlm.nih.gov/pubmed/29852428

[111] Chang X, Li H, Feng J, Chen Y, Nie G, Zhang J. Effects of cadmium exposure on the composition and diversity of the intestinal microbial community of common carp (*Cyprinus carpio* L.). Ecotoxicology and Environmental Safety. 2019;**171**:92-98. Available from: http://www.ncbi.nlm.nih.gov/pubmed/30597321

[112] Kalantzi I, Zeri C, Catsiki V-A, Tsangaris C, Strogyloudi E, Kaberi H, et al. Assessment of the use of copper alloy aquaculture nets: Potential impacts on the marine environment and on the farmed fish. Aquaculture. 2016;**465**:209-222. Available from: https://www.sciencedirect.com/science/article/pii/S0044848616304690

[113] Pradel P, Corsini G, Tello M, González A. Pantoea agglomerans an agent to remove residual copper from aquaculture activity. Advances in Materials Research. 2014;**945-949**:3479-3482

[114] Davidson J, Good C, Welsh C, Summerfelt ST. Abnormal swimming behavior and increased deformities in rainbow trout *Oncorhynchus mykiss* cultured in low exchange water recirculating aquaculture systems. Aquacultural Engineering. 2011;**45**(3):109-117. Available from: https://www.sciencedirect.com/science/article/pii/S014486091100063X

[115] Rud I, Kolarevic J, Holan AB, Berget I, Calabrese S, Terjesen BF. Deep-sequencing of the bacterial microbiota in commercial-scale recirculating and semi-closed aquaculture systems for Atlantic salmon post-smolt production. Aquacultural Engineering. 2017;**78**:50-62. Available from: https://www.sciencedirect.com/science/article/pii/S0144860916301480

[116] Aguilera E, Yany G, Romero J. Cultivable intestinal microbiota of Seriola lalandi cultivable intestinal microbiota of yellowtail juveniles (*Seriola lalandi*) in an aquaculture system. Latin American Journal of Aquatic Research. 2013;**41**(3):395-403. Available from: http://www.fao.org/fishery/statistics/software/

[117] Salas Leiva J, Opazo R, Remond C, Uribe E, Velez A, Romero J. Characterization of the intestinal microbiota of wild caught and farmed fine flounder (*Paralichthys adspersus*). Latin American Journal of Aquatic Research. 2017;**45**(2):370-378. Available from: http://www.lajar.cl/pdf/imar/v45n2/Artículo_45_2_12.pdf

[118] Ramírez C, Romero J. The microbiome of Seriola lalandi of wild and aquaculture origin reveals differences in composition and potential function. Frontiers in Microbiology. 2017;**8**:1844. Available from: http://www.ncbi.nlm.nih.gov/pubmed/29018423

[119] Ramírez C, Romero J. Fine flounder (*Paralichthys adspersus*) microbiome showed important differences between wild and reared specimens. Frontiers in Microbiology. 2017;**8**:271. Available from: http://www.ncbi.nlm.nih.gov/pubmed/28286497

[120] Navarrete P, Espejo RT, Romero J. Molecular analysis of microbiota along

the digestive tract of juvenile Atlantic salmon (Salmo salar L.). Microbial Ecology. 2009;**57**(3):550-561. Available from: http://link.springer.com/10.1007/s00248-008-9448-x

[121] Navarrete P, Magne F, Araneda C, Fuentes P, Barros L, Opazo R, et al. PCR-TTGE analysis of 16S rRNA from rainbow trout (Oncorhynchus mykiss) gut microbiota reveals host-specific communities of active bacteria. PLoS One. 2012;**7**(2):e31335. Available from: http://www.pubmedcentral.nih.gov/articlerender.fcgi?artid=3290605&tool=pmcentrez&rendertype=abstract

[122] Cárcamo C, Díaz N, Winkler F. Genetic diversity in Chilean populations of rainbow trout, *Oncorhynchus mykiss*. Latin American Journal of Aquatic Research. 2015;**43**(1):49. Available from: http://www.lajar.cl/resumen_en.php?cod=20150310135000&id=67

[123] Romero J, Navarrete P. 16S rDNA-based analysis of dominant bacterial populations associated with early life stages of Coho Salmon (*Oncorhynchus kisutch*). Microbial Ecology. 2006;**51**(4):422-430. Available from: http://www.ncbi.nlm.nih.gov/pubmed/16598631

[124] Chiarello M, Auguet J-C, Bettarel Y, Bouvier C, Claverie T, Graham NAJ, et al. Skin microbiome of coral reef fish is highly variable and driven by host phylogeny and diet. Microbiome. 2018;**6**(1):147. Available from: http://www.ncbi.nlm.nih.gov/pubmed/30143055

[125] Littman DR, Pamer EG. Role of the commensal microbiota in Normal and pathogenic host immune responses. Cell Host & Microbe. 2011;**10**(4):311-323. Available from: http://www.ncbi.nlm.nih.gov/pubmed/22018232

[126] Pankhurst NW, King HR. Temperature and salmonid reproduction: Implications for aquaculture. Journal of Fish Biology. 2010;**76**(1):69-85. Available from: http://www.ncbi.nlm.nih.gov/pubmed/20738700

[127] Martínez Cruz P, Ibáñez AL, Monroy Hermosillo OA, Ramírez Saad HC. Use of probiotics in aquaculture. ISRN Microbiology. 2012;**2012**:916845. Available from: http://www.ncbi.nlm.nih.gov/pubmed/23762761

[128] Vaughan EE, de Vries MC, Zoetendal EG, Ben-Amor K, Akkermans ADL, de Vos WM. The intestinal LABs. Antonie Van Leeuwenhoek. 2002;**82**(1-4):341-352. Available from: http://www.ncbi.nlm.nih.gov/pubmed/12369201

[129] Balcázar JL, Blas I de, Ruiz-Zarzuela I, Cunningham D, Vendrell D, Múzquiz JL. The role of probiotics in aquaculture. Veterinary Microbiology 2006;**114**(3-4):173-186. Available from: https://www.sciencedirect.com/science/article/pii/S0378113506000265

[130] Lazado CC, Caipang CMA, Brinchmann MF, Kiron V. In vitro adherence of two candidate probiotics from Atlantic cod and their interference with the adhesion of two pathogenic bacteria. Veterinary Microbiology. 2011;**148**(2-4):252-259. Available from: http://www.ncbi.nlm.nih.gov/pubmed/20884135

[131] Lazado CC, Caipang CMA, Estante EG. Prospects of host-associated microorganisms in fish and penaeids as probiotics with immunomodulatory functions. Fish & Shellfish Immunology. 2015;**45**(1):2-12. Available from: http://www.ncbi.nlm.nih.gov/pubmed/25703713

[132] Nayak SK. Probiotics and immunity: A fish perspective. Fish & Shellfish Immunology. 2010;**29**(1):2-14. Available from: http://www.ncbi.nlm.nih.gov/pubmed/20219683

[133] Akhter N, Wu B, Memon AM, Mohsin M. Probiotics and prebiotics associated with aquaculture: A review. Fish & Shellfish Immunology. 2015;**45**(2):733-741. Available from: http://www.ncbi.nlm.nih.gov/pubmed/26044743

[134] Verschuere L, Rombaut G, Sorgeloos P, Verstraete W. Probiotic bacteria as biological control agents in aquaculture. Microbiology and Molecular Biology Reviews. 2000;**64**(4):655-671. Available from: http://www.ncbi.nlm.nih.gov/pubmed/11104813

[135] De Schryver P, Sinha AK, Kunwar PS, Baruah K, Verstraete W, Boon N, et al. Poly-β-hydroxybutyrate (PHB) increases growth performance and intestinal bacterial range-weighted richness in juvenile European sea bass, Dicentrarchus labrax. Applied Microbiology and Biotechnology. 2010;**86**(5):1535-1541. Available from: http://www.ncbi.nlm.nih.gov/pubmed/20094715

[136] Nhan DT, Wille M, De Schryver P, Defoirdt T, Bossier P, Sorgeloos P. The effect of poly β-hydroxybutyrate on larviculture of the giant freshwater prawn Macrobrachium rosenbergii. Aquaculture. 2010;**302**(1-2):76-81. Available from: https://www.sciencedirect.com/science/article/abs/pii/S0044848610000992

[137] Najdegerami EH, Baruah K, Shiri A, Rekecki A, Van den Broeck W, Sorgeloos P, et al. Siberian sturgeon (Acipenser baerii) larvae fed Artemia nauplii enriched with poly-β-hydroxybutyrate (PHB): Effect on growth performance, body composition, digestive enzymes, gut microbial community, gut histology and stress tests. Aquaculture Research. 2015;**46**(4):801-812. Available from: http://doi.wiley.com/10.1111/are.12231

[138] Sui L, Ma G, Lu W, Deng Y, Bossier P, De Schryver P. Effect of poly-β-hydroxybutyrate on growth, enzyme activity and intestinal microbial community of Chinese mitten crab, Eriocheir sinensis (Milne-Edwards) juveniles. Aquaculture Research. 2016;**47**(11):3644-3652. Available from: http://doi.wiley.com/10.1111/are.12817

[139] Ludevese-Pascual G, Laranja JLQ, Amar EC, Sorgeloos P, Bossier P, De Schryver P. Poly-beta-hydroxybutyrate-enriched Artemia sp. for giant tiger prawn Penaeus monodon larviculture. Aquaculture Nutrition. 2017;**23**(2):422-429. Available from: http://doi.wiley.com/10.1111/anu.12410

[140] Henríquez Parada CP. Caracterización de propiedades probióticas de microorganismos del tracto digestivo de salmónidos. 2013. Available from: http://repositorio.uchile.cl/handle/2250/116254

[141] Riquelme C, Hayashida G, Araya R, Uchida A, Satomi M, Ishida Y. Isolation of a native bacterial strain the scallop Argopecten purpuratus with inhibitory effects against pathogenic vibrios. Journal of Shellfish Research. 1996;**15**(2):369-374. Available from: https://www.researchgate.net/publication/279674783

[142] Riquelme C, Araya R, Vergara N, Rojas A, Guaita M, Candia M. Potential probiotic strains in the culture of the Chilean scallop Argopecten purpuratus (Lamarck, 1819). Aquaculture. 1997;**154**(1):17-26. Available from: https://www.sciencedirect.com/science/article/pii/S0044848697000434

[143] Van Hung N, De Schryver P, Dung NV, Nevejan N, Bossier P. Ralstonia eutropha, containing high poly-β-hydroxybutyrate levels, regulates the immune response in mussel larvae challenged with vibrio coralliilyticus. Fish & Shellfish Immunology. 2019;**84**:196-203. Available from: https://www.sciencedirect.com/science/article/pii/S1050464818306107

[144] Gibson L, Woodworth J, George A. Probiotic activity of Aeromonas media on the Pacific oyster, Crassostrea gigas, when challenged with vibrio tubiashii. Aquaculture. 1998;**169**(1-2): 111-120. Available from: https://www.sciencedirect.com/science/article/pii/S004484869800369X

[145] Gould AL, Zhang V, Lamberti L, Jones EW, Obadia B, Korasidis N, et al. Microbiome interactions shape host fitness. Proceedings of the National Academy of Sciences. 2018;**115**(51):E11951-E11960. Available from: http://www.ncbi.nlm.nih.gov/pubmed/30510004

[146] Langille MGI, Zaneveld J, Caporaso JG, McDonald D, Knights D, Reyes JA, et al. Predictive functional profiling of microbial communities using 16S rRNA marker gene sequences. Nature Biotechnology. 2013;**31**(9):814-821. Available from: http://www.nature.com/articles/nbt.2676

[147] Blin K, Wolf T, Chevrette MG, Lu X, Schwalen CJ, Kautsar SA, et al. antiSMASH 4.0—Improvements in chemistry prediction and gene cluster boundary identification. Nucleic Acids Research. 2017;**45**(W1):W36-W41. Available from: https://academic.oup.com/nar/article-lookup/doi/10.1093/nar/gkx319

[148] Geuking MB, Köller Y, Rupp S, McCoy KD. The interplay between the gut microbiota and the immune system. Gut Microbes. 2016;**5**(3):411-418. Available from: http://www.ncbi.nlm.nih.gov/pubmed/24922519

[149] Chu H, Mazmanian SK. Innate immune recognition of the microbiota promotes host-microbial symbiosis. Nature Immunology. 2013;**14**(7): 668-675. DOI: 10.1038/ni.2635

[150] Wlodarska M, Thaiss CA, Nowarski R, Henao-Mejia J, Zhang J-P, Brown EM, et al. NLRP6 inflammasome orchestrates the colonic host-microbial interface by regulating goblet cell mucus secretion. Cell. 2014;**156**(5): 1045-1059. Available from: http://www.pubmedcentral.nih.gov/articlerender.fcgi?artid=4017640&tool=pmcentrez&rendertype=abstract

[151] Kawashima T, Kosaka A, Yan H, Guo Z, Uchiyama R, Fukui R, et al. Double-stranded RNA of intestinal commensal but not pathogenic bacteria triggers production of protective interferon-β. Immunity. 2013;**38**(6):1187-1197. Available from: http://www.ncbi.nlm.nih.gov/pubmed/23791646

[152] Abt MC, Osborne LC, Monticelli LA, Doering TA, Alenghat T, Sonnenberg GF, et al. Commensal bacteria calibrate the activation threshold of innate antiviral immunity. Immunity. 2012;**37**(1):158-170. Available from: http://www.pubmedcentral.nih.gov/articlerender.fcgi?artid=3679670&tool=pmcentrez&rendertype=abstract

[153] Ichinohe T, Pang IK, Kumamoto Y, Peaper DR, Ho JH, Murray TS, et al. Microbiota regulates immune defense against respiratory tract influenza A virus infection. Proceedings of the National Academy of Sciences of the United States of America. 2011;**108**(13):5354-5359. Available from: http://www.pubmedcentral.nih.gov/articlerender.fcgi?artid=3069176&tool=pmcentrez&rendertype=abstract

[154] Round JL, Mazmanian SK. Inducible Foxp3+ regulatory T-cell development by a commensal bacterium of the intestinal microbiota. Proceedings of the National Academy of Sciences of the United States of America. 2010;**107**(27):12204-12209. Available from: http://www.pubmedcentral.nih.gov/articlerender.fcgi?artid=2901479&tool=pmcentrez&rendertype=abstract

[155] Round JL, Lee SM, Li J, Tran G, Jabri B, Chatila TA, et al. The

toll-like receptor 2 pathway establishes colonization by a commensal of the human microbiota. Science. 2011;**332**(6032):974-977. Available from: http://www.sciencemag.org/content/332/6032/974.short

[156] Rauta PR, Samanta M, Dash HR, Nayak B, Das S. Toll-like receptors (TLRs) in aquatic animals: Signaling pathways, expressions and immune responses. Immunology Letters. 2014;**158**(1-2):14-24. Available from: http://www.ncbi.nlm.nih.gov/pubmed/24291116

[157] Pietretti D, Wiegertjes GF. Ligand specificities of toll-like receptors in fish: Indications from infection studies. Developmental and Comparative Immunology. 2014;**43**(2):205-222. Available from: http://www.ncbi.nlm.nih.gov/pubmed/23981328

[158] Salazar C, Haussmann D, Kausel G, Figueroa J. Molecular cloning of Salmo salar toll-like receptors (TLR1, TLR22, TLR5M and TLR5S) and expression analysis in SHK-1 cells during Piscirickettsia salmonis infection. Journal of Fish Diseases. 2016;**39**(2):239-248. Available from: http://www.ncbi.nlm.nih.gov/pubmed/25903926

[159] Ost KS, Round JL. Communication between the microbiota and mammalian immunity. Annual Review of Microbiology. 2018;**72**(1):399-422. Available from: http://www.ncbi.nlm.nih.gov/pubmed/29927706

[160] Cummings JH, Pomare EW, Branch WJ, Naylor CP, Macfarlane GT. Short chain fatty acids in human large intestine, portal, hepatic and venous blood. Gut. 1987;**28**(10):1221-1227. Available from: http://www.pubmedcentral.nih.gov/articlerender.fcgi?artid=1433442&tool=pmcentrez&rendertype=abstract

[161] Arpaia N, Campbell C, Fan X, Dikiy S, van der Veeken J, deRoos P, et al. Metabolites produced by commensal bacteria promote peripheral regulatory T-cell generation. Nature. 2013;**504**(7480):451-455. Available from: http://www.pubmedcentral.nih.gov/articlerender.fcgi?artid=3869884&tool=pmcentrez&rendertype=abstract

[162] Le Poul E, Loison C, Struyf S, Springael J-Y, Lannoy V, Decobecq M-E, et al. Functional characterization of human receptors for short chain fatty acids and their role in polymorphonuclear cell activation. The Journal of Biological Chemistry. 2003;**278**(28):25481-25489. Available from: http://www.ncbi.nlm.nih.gov/pubmed/12711604

[163] Erny D, Hrabě de Angelis AL, Jaitin D, Wieghofer P, Staszewski O, David E, et al. Host microbiota constantly control maturation and function of microglia in the CNS. Nature Neuroscience. 2015;**18**(7):965-977. Available from: http://www.ncbi.nlm.nih.gov/pubmed/26030851

[164] Cox MA, Jackson J, Stanton M, Rojas-Triana A, Bober L, Laverty M, et al. Short-chain fatty acids act as antiinflammatory mediators by regulating prostaglandin E(2) and cytokines. World Journal of Gastroenterology. 2009;**15**(44):5549-5557. Available from: http://www.pubmedcentral.nih.gov/articlerender.fcgi?artid=2785057&tool=pmcentrez&rendertype=abstract

[165] Eftimiadi C, Stashenko P, Tonetti M, Mangiante PE, Massara R, Zupo S, et al. Divergent effect of the anaerobic bacteria by-product butyric acid on the immune response: Suppression of T-lymphocyte proliferation and stimulation of interleukin-1 beta production. Oral Microbiology and Immunology. 1991;**6**(1):17-23. Available from: http://www.ncbi.nlm.nih.gov/pubmed/1945479

[166] Zimmerman MA, Singh N, Martin PM, Thangaraju M, Ganapathy V,

Waller JL, et al. Butyrate suppresses colonic inflammation through HDAC1-dependent Fas upregulation and Fas-mediated apoptosis of T cells. American Journal of Physiology. Gastrointestinal and Liver Physiology. 2012;**302**(12):G1405-G1415. Available from: http://www.pubmedcentral.nih.gov/articlerender.fcgi?artid=3378095&tool=pmcentrez&rendertype=abstract

[167] Berndt BE, Zhang M, Owyang SY, Cole TS, Wang TW, Luther J, et al. Butyrate increases IL-23 production by stimulated dendritic cells. American Journal of Physiology. Gastrointestinal and Liver Physiology. 2012;**303**(12):G1384-G1392. Available from: http://www.pubmedcentral.nih.gov/articlerender.fcgi?artid=3532546&tool=pmcentrez&rendertype=abstract

[168] Liu L, Li L, Min J, Wang J, Wu H, Zeng Y, et al. Butyrate interferes with the differentiation and function of human monocyte-derived dendritic cells. Cellular Immunology. 2015;**277**(1-2):66-73. Available from: http://www.ncbi.nlm.nih.gov/pubmed/22698927

[169] Hinnebusch BF, Meng S, Wu JT, Archer SY, Hodin RA. The effects of short-chain fatty acids on human colon cancer cell phenotype are associated with histone hyperacetylation. The Journal of Nutrition. 2002;**132**(5):1012-1017. Available from: http://www.ncbi.nlm.nih.gov/pubmed/11983830

[170] Waldecker M, Kautenburger T, Daumann H, Busch C, Schrenk D. Inhibition of histone-deacetylase activity by short-chain fatty acids and some polyphenol metabolites formed in the colon. The Journal of Nutritional Biochemistry. 2008;**19**(9):587-593. Available from: http://www.ncbi.nlm.nih.gov/pubmed/18061431

[171] Holben WE, Williams P, Gilbert MA, Saarinen M, Särkilahti LK, Apajalahti JHA. Phylogenetic analysis of intestinal microflora indicates a novel mycoplasma phylotype in farmed and wild salmon. Microbial Ecology. 2002;**44**(2):175-185. Available from: http://www.ncbi.nlm.nih.gov/pubmed/12082453

[172] Mountfort DO, Campbell J, Clements KD. Hindgut fermentation in three species of marine herbivorous fish. Applied and Environmental Microbiology. 2002;**68**(3):1374-1380. Available from: http://www.pubmedcentral.nih.gov/articlerender.fcgi?artid=123746&tool=pmcentrez&rendertype=abstract

[173] Robles R, Lozano AB, Sevilla A, Márquez L, Nuez-Ortín W, Moyano FJ. Effect of partially protected butyrate used as feed additive on growth and intestinal metabolism in sea bream (*Sparus aurata*). Fish Physiology and Biochemistry. 2013;**39**(6):1567-1580. Available from: http://www.ncbi.nlm.nih.gov/pubmed/23737146

[174] Liu W, Yang Y, Zhang J, Gatlin DM, Ringø E, Zhou Z. Effects of dietary microencapsulated sodium butyrate on growth, intestinal mucosal morphology, immune response and adhesive bacteria in juvenile common carp (*Cyprinus carpio*) pre-fed with or without oxidised oil. The British Journal of Nutrition. 2014;**112**(1):15-29. Available from: http://www.ncbi.nlm.nih.gov/pubmed/24774835

[175] Kumar H, Lund R, Laiho A, Lundelin K, Ley RE, Isolauri E, et al. Gut microbiota as an epigenetic regulator: Pilot study based on whole-genome methylation analysis. MBio. 2014;**5**(6):e02113-14. Available from: http://www.pubmedcentral.nih.gov/articlerender.fcgi?artid=4271550&tool=pmcentrez&rendertype=abstract

[176] Berger SL. The complex language of chromatin regulation during transcription. Nature. 2007;**447**(7143):407-412. Available

from: http://www.ncbi.nlm.nih.gov/pubmed/17522673

[177] Sabari BR, Zhang D, Allis CD, Zhao Y. Metabolic regulation of gene expression through histone acylations. Nature Reviews. Molecular Cell Biology. 2016;**18**(2):90-101. Available from: http://www.ncbi.nlm.nih.gov/pubmed/27924077

[178] Duval M, Cossart P, Lebreton A. Mammalian microRNAs and long noncoding RNAs in the host-bacterial pathogen crosstalk. Seminars in Cell & Developmental Biology. 2016;**65**:11-19

[179] Qin Y, Wade PA. Crosstalk between the microbiome and epigenome: Messages from bugs. Journal of Biochemistry. 2018;**163**(2):105-112. Available from: http://www.ncbi.nlm.nih.gov/pubmed/29161429

[180] Obata Y, Furusawa Y, Endo TA, Sharif J, Takahashi D, Atarashi K, et al. The epigenetic regulator Uhrf1 facilitates the proliferation and maturation of colonic regulatory T cells. Nature Immunology. 2014;**15**(6):571-579. Available from: http://www.ncbi.nlm.nih.gov/pubmed/24777532

[181] Ganal SC, Sanos SL, Kallfass C, Oberle K, Johner C, Kirschning C, et al. Priming of natural killer cells by nonmucosal mononuclear phagocytes requires instructive signals from commensal microbiota. Immunity. 2012;**37**(1):171-186. Available from: http://www.ncbi.nlm.nih.gov/pubmed/22749822

[182] Chang PV, Hao L, Offermanns S, Medzhitov R. The microbial metabolite butyrate regulates intestinal macrophage function via histone deacetylase inhibition. Proceedings of the National Academy of Sciences of the United States of America. 2014;**111**(6):2247-2252. Available from: http://www.pubmedcentral.nih.gov/articlerender.fcgi?artid=3926023&tool=pmcentrez&rendertype=abstract

[183] Alenghat T, Osborne LC, Saenz SA, Kobuley D, Ziegler CGK, Mullican SE, et al. Histone deacetylase 3 coordinates commensal-bacteria-dependent intestinal homeostasis. Nature. 2013;**504**(7478):153-157. Available from: http://www.pubmedcentral.nih.gov/articlerender.fcgi?artid=3949438&tool=pmcentrez&rendertype=abstract

[184] Best C, Ikert H, Kostyniuk DJ, Craig PM, Navarro-Martin L, Marandel L, et al. Epigenetics in teleost fish: From molecular mechanisms to physiological phenotypes. Comparative Biochemistry and Physiology Part B: Biochemistry and Molecular Biology. 2018;**224**:210-244. Available from: http://www.ncbi.nlm.nih.gov/pubmed/29369794

[185] Liang L, Ai L, Qian J, Fang J-Y, Xu J, Lakhdari O, et al. Long noncoding RNA expression profiles in gut tissues constitute molecular signatures that reflect the types of microbes. Scientific Reports. 2015;**5**:11763. Available from: http://www.nature.com/articles/srep11763

[186] Masotti A. Interplays between gut microbiota and gene expression regulation by miRNAs. Frontiers in Cellular and Infection Microbiology. 2012;**2**:137. Available from: http://www.ncbi.nlm.nih.gov/pubmed/23130352

[187] Liang L, Ai L, Qian J, Fang J-Y, Xu J. Long noncoding RNA expression profiles in gut tissues constitute molecular signatures that reflect the types of microbes. Scientific Reports. 2015;**5**:11763. Available from: http://www.ncbi.nlm.nih.gov/pubmed/26123364

[188] Taganov KD, Boldin MP, Chang K-J, Baltimore D. NF-kappaB-dependent induction of microRNA miR-146, an inhibitor targeted to signaling proteins of innate immune responses.

Proceedings of the National Academy of Sciences of the United States of America. 2006;**103**(33):12481-12486. Available from: http://www.ncbi.nlm.nih.gov/pubmed/16885212

[189] Schulte LN, Westermann AJ, Vogel J. Differential activation and functional specialization of miR-146 and miR-155 in innate immune sensing. Nucleic Acids Research. 2013;**41**(1):542-553. Available from: http://www.ncbi.nlm.nih.gov/pubmed/23143100

[190] Duerr CU, Hornef MW. The mammalian intestinal epithelium as integral player in the establishment and maintenance of host–microbial homeostasis. Seminars in Immunology. 2012;**24**(1):25-35

[191] Andreassen R, Worren M, Høyheim B, Bartel D, Chekulaeva M, Filipowicz W, et al. Discovery and characterization of miRNA genes in Atlantic salmon (*Salmo salar*) by use of a deep sequencing approach. BMC Genomics. 2013;**14**(1):482. Available from: http://bmcgenomics.biomedcentral.com/articles/10.1186/1471-2164-14-482

[192] Food and Agriculture Organization of the United Nations. Aquaculture development. 4, Ecosystem approach to aquaculture. Food and Agriculture Organization of the United Nations. 2010;**5**(suppl 4):53. Available from: http://www.fao.org/docrep/013/i1750e/i1750e00.htm

[193] León-Muñoz J, Urbina MA, Garreaud R, Iriarte JL. Hydroclimatic conditions trigger record harmful algal bloom in western Patagonia (summer 2016). Scientific Reports. 2018;**8**(1):1330. Available from: http://www.nature.com/articles/s41598-018-19461-4

[194] Seong KA, Jeong HJ. Interactions between the pathogenic bacterium Vibrio parahaemolyticus and red-tide dinoflagellates. Ocean Science Journal. 2011;**46**(2):105-115. Available from: http://link.springer.com/10.1007/s12601-011-0010-2

[195] Imai I, Kimura S. Resistance of the fish-killing dinoflagellate Cochlodinium polykrikoides against algicidal bacteria isolated from the coastal sea of Japan. Harmful Algae. 2008;**7**(3):360-367. Available from: https://www.sciencedirect.com/science/article/pii/S1568988307001837

[196] Imai I, Sunahara T, Nishikawa T, Hori Y, Kondo R, Hiroishi S. Fluctuations of the red tide flagellates Chattonella spp. (Raphidophyceae) and the algicidal bacterium Cytophaga sp. in the Seto Inland Sea, Japan. Marine Biology. 2001;**138**(5):1043-1049. Available from: http://link.springer.com/10.1007/s002270000513

[197] Imai I, Ishida Y, Sakaguchi K, Hata Y. Algicidal marine bacteria isolated from northern Hiroshima Bay, Japan. Fisheries Science. 1995;**61**(4):628-636. Available from: https://www.jstage.jst.go.jp/article/fishsci1994/61/4/61_4_628/_article

[198] Imai I, Ishida Y, Hata Y. Killing of marine phytoplankton by a gliding bacterium Cytophaga sp., isolated from the coastal sea of Japan. Marine Biology. 1993;**116**(4):527-532. Available from: http://link.springer.com/10.1007/BF00355470